Tillandsia

The World's Most Unusual Airplants

Tillandsia

The World's Most Unusual Air Plants

By Paul T. Isley III

BOTANICAL PRESS

Gardena, California 90249

1987

The Bromeliad Society, Inc. is a horticultural organization that was formed in 1950 to promote interest in and disseminate knowledge about this fascinating family of plants. With members from dozens of countries around the globe, the Society caters to apartment dwellers with an interest in house plants, gardening enthusiasts, greenhouse hobbyists, nurserymen, teachers, scientists, and botanical garden personnel.

All persons who are interested in bromeliads, in learning more about them, in growing them, and in participating in the activities of The Bromeliad Society are welcome to become members. For more information write to The Bromeliad Society, Inc., 2488 East 49th, Tulsa, Oklahoma, 74105, U.S.A.

Library of Congress Catalog Card Number: 86-072858

ISBN: 0-9617675-0-2

Published by: Botanical Press
P.O. Box 5592
Gardena, California 90291
(213) 515-5200

Book preparation by Rosa Schuth
Printed in Japan by Dai Nippon Printing Company.

Library of Congress Cataloging in Publication Data

Isley, Paul T.
Tillandsia; the world's most unusual airplants.

Includes bibliographies and index.
1. Tillandsia. 2. Tillandsia—Classification.
I. Title.
SB413.T52I65 1987 635.9'3422 86-72858
ISBN 0-9617675-0-2

Dedicated To
the Memory of Mary Wisdom

Acknowledgments

Without the initial support and enthusiasm of Mary Wisdom of New Orleans, this book would not have been possible. Over a span of 25 years, she developed a keen appreciation for bromeliads in general and tillandsias in particular. Her system of elaborate notes grew to fill many binders. She was well aware of the need for a book about her favorite plants.

As I look back on the three years spent taking the photographs and preparing the manuscript (unless otherwise noted, all photographs are by the author), many helpful faces come to mind. First and foremost, I would like to thank David Benzing of Oberlin College. His ideas, expertise, and, above all, patience, were critical to the initial development of the manuscript. His book, *The Biology of Bromeliads,* is remarkable. It provides detailed information on a broad range of related subjects. If *Tillandsia* kindles a fire to learn more about the plants, his book is a "must."

I would also like to thank the late Victoria Padilla, Hal Wiedman of California State University at Sacramento, Amy Jean Gilmartin of Washington State University, Gregory Brown of the University of Wyoming, Harry Luther of the Marie Selby Botanical Gardens in Sarasota, Pat Morishita of the University of California at Riverside (retired), Laurence Nobles of Northwestern University, and Robert Smaus and Linda French of the Los Angeles Times Magazine for critically reading parts of the manuscript.

Lyman B. Smith and Robert Read of the National Herbarium at the Smithsonian Institution assisted me in my studies at the herbarium. The librarian, Ruth Schallert, gave freely of her valuable time and helped with the biographical sketches. My thanks and appreciation are due to James White of the Hunt Institute for Botanical Documentation at the Carnegie-Mellon University. Racine Foster supplied the photograph and background material of her late husband, Mulford. Werner Rauh and his staff at the University of Heidelberg in West Germany availed themselves to the project. Rauh described *Tillandsia muhrii* at my request so that it could be included in the book. The superb scanning electron microscope (SEM) photographs of species' trichomes were produced by his assistant, Christine Dannenbaum. They contribute immensely and will intrigue all who see them—as will figures 6-18 through 16-21 which were taken by Elvira Gross.

This book is the culmination of 20 years involvement with *Tillandsia.* My good friend Mark Dimmitt of the Arizona-Sonora Desert Museum in Tucson gave me my first "little marbles" and encouraged my interest. William Harris (1927-1985) of Guatemala began shipping me plants in 1974. We "hit it off" immediately, and he became my business mentor. Soon thereafter we formed Rainforest Flora, Inc. His knowledge, experience, sense of humor, and, above all, his wisdom were an inspiration to me. Don Bleitz (1921-1986) was another man whose life, business experience, and friendship meant a great deal both to me and to the completion of this project.

The two people who gave the most support in the production of this work were my wife, Elizabeth, and my compatriot and business partner, Jerrold Robinson. Betty gave me great support and patience without complaint. There is no way I can express in words my appreciation and love for her. Jerry and I have been through some incredible times together since we first met coincidentally on a Caribbean diving expedition for sunken treasure in the early '70s. This book would not exist without his constant dedication—which has always been far above the norm.

There are many others, including Sue Gardner, Craig Martin, Father Charles Cavanach, S.J., and Steve Thornton who have volunteered their time and expertise. With great technical execution, Tina Ross has done a marvelous job with the drawings, adding significantly to their development and interpretation. Gloria Simmons and Dena Phillips at Digital Typographers are patient professionals. The representative for Dai Nippon Printing Company in America, Kohei Tsumori, was knowledgable, diligent, and honest—the perfect liason between myself and the company. Thank you one and all.

Final responsibility for the writing, photography, development, and production of the book rests on my shoulders. This, of course, includes any and all mistakes and/or inaccuracies.

Vivian Mercier in his *The Irish Comic Tradition* accurately stated my sentiments when he said, "This book makes no claim to be the last word on its subject: it is much closer to being the first one."

Grateful acknowledgment is made to the following for material used or permission to reproduce material. Figures 3.1, 3.2, 3.7, 3.16, 3.18, and 3.19 are courtesy of The Hunt Institute for Botanical Documentation, Carnegie-Mellon University, Pittsburgh, Pennsylvania.

The pronounciation chart on page 172 is from *Botanical Latin* by William T. Stearn and is used with the permission of the publisher, David & Charles Ltd., Brunel House, Newton Abbot Devon, England, TQ12 4PU. This book is fascinating and highly recommended for those with a desire to learn more about the historical development of botanical Latin and its grammatical use, syntax, punctuation, and vocabulary. The book also devotes space to the methods for describing species.

The color chart on page 241 is used with the permission of Dr. A.B. Graf, author of *Exotica,* Pictorial Cyclopedia of Exotic Plants, published by Roehrs Company, East Rutherford, New Jersey, 07073.

Figure 1.6 is courtesy of Michael Parella, of the University of California at Riverside. Figures 6-27, 6-28, 6-30, and 6-31 are courtesy of Greg Brown.

Preface

As the popularity of *Tillandsia* has increased, so has the need for a book that not only presents the plants in bloom but educates the public as well. Without a means to learn about *Tillandsia* in general, and various species in particular, a person cannot achieve a high level of knowledge and appreciation. This book is a major step toward that goal.

I have tried to produce a book about *Tillandsia* that I would like to see and read. Subjects such as taxonomy, etymology, biographical histories, plant biology, and evolution have fascinated me for years.

When I began this project, my intent was to produce a soft cover book that would retail for about $12. However, try as I might, I could not develop a format that would come close to what I was trying to achieve—namely, a work with photos and text of enough species to give the reader a solid base of knowledge and appreciation for *Tillandsia.* As I went along the pendulum swung in the other direction toward a more complete treatment.

Tillandsia has been crafted of the finest materials with an eye toward providing a proper setting in which to display the plants in a book that will endure in good condition for many years. The Artist Paper is the highest quality, glossy stock available. It is heavy and opaque so that ink will not bleed through to the back side of a page. The layout has been designed with an open format that allows space for the photographs to be appreciated individually. The dust cover is french folded to resist tearing. The smyth sown binding and hard cover assure that frequent use will not adversely affect the book itself.

The meager supply of current literature tends to concentrate either on works that are purely scientific or on those that use a photographic format with abbreviated text. A technical treatise is adequate if one is writing a book for botanists and making a scientific contribution. In such a book, beautiful color photographs are not necessary. On the other hand, a book of photographs is relatively easy to produce and makes a great visual impact—but it contains little information.

Tillandsia combines high-quality color photographs with enough information to educate the curious. The text contains a fair bit of jargon, but technical words have a useful purpose because they condense ideas. Most of the terms are defined in the text the first time they are used. The glossary provides sufficient assistance, the bookmark a convenient aid.

For those who want merely a casual understanding of the plants, the text is arranged so that botanical terms can be skipped while the sense of their meanings remains.

Choosing the tillandsias presented in this book was not easy because there are so many worthy choices. The final decision was based on informal criteria that I felt would most enhance the book, the plants, and the interested public. By and large, I picked species that were visually interesting, available, hardy, and relatively inexpensive.

In the chapter on Species, the taxonomist who named the plant and the year the description was first published is printed on the line below the species name. The Latin pronounciation is under that; the *Tillandsia* subgenus is on the next line. The etymology is printed in bold-face type in the first paragraph of the description.

Throughout the book, a *Tillandsia* mentioned in text that is not featured in the Species chapter has the taxonomist's name listed immediately after the epithet the first time it is used.

In discussing care conditions for the various species, I frequently made reference to "average care conditions." This means care relative to other tillandsias. There are as many growing environments as there are growers. But within a plant's *successful* habitat there are "average" conditions in terms of light, temperature, humidity, frequency of watering and fertilizing. These facets are covered in the introductory sections on Outdoor Care and Indoor Care and the chapter on *Tillandsia* Biology.

Evolution and Biology were the most difficult subjects about which to write. They are inherently complex and technical, and a thorough treatment of current fact and theory would fill a large scientific volume and would assume a botanist's background. Still, tillandsias are so variable and flourish in such a wide range of habitats that for years I have been curious as to why and how they were able to do it.

Many readers studied biology in high school or college but have forgotten much of that material. With a keen interest in *Tillandsia,* there will now be inspiration to learn something more about water and food transport, photosynthesis, the blooming cycle, the trichomes, and so on.

Because it was difficult to condense these subjects into a few informational pages, I had to assume that the reader would be familiar with a number of biological and/or genetic topics. Use the glossary and/or a high school biology book to fill in the gaps.

At this time, tillandsias are not widely available. If you cannot find tillandsias, ask the people at your local nursery to order them, or write to Rainforest Flora, Inc., 1927 W. Rosecrans Ave., Gardena, CA 90249, which supplies plants wholesale, retail, and by mail order.

PTI III

Contents

Among the plants which were my favorite objects during the long trek across the Cordilleras were the Bromeliaceae. For fifteen years they have been the target of my increasing enthusiasm. . . Gradually a whole group of botanists and horticulturists was captivated by the Bromeliaceae because of. . . the beauty of their foliage, their regular and strange mode of growth, their often brilliant flowers which are sometimes ephemeral, other times lasting for a long period, their decorative value for the greenhouse and the home, the ease with which they can be cultivated, etc.

Large species such as *T. secunda (T. minima)* and *T. paniculata* take hold in the tops of large trees, often overtopping them by as much as 3 meters or more with their scapes, jutting up as grandiose flowering candelabras. The weakest of branches are adequate for supporting them and they are able to keep erect due to an unfailing balance. You might ask yourself how they are able to withstand the gusts of winds if you did not remember that there are hardly any winds within this equatorial territory where the ancient *conquistadores* discovered the famous "Pacific" Ocean and which Humboldt called correctly "the region of calms."

The green luster of the *Tillandsia* leaves is often very brilliant and shimmering in the sun when it is not richly marked with purple, wine, or blood, spotted or marked with zebra stripes. This coloration of the Bromeliaceae is so striking that several of them are appreciated by amatures only because of the beautiful leaves.

More than 700 species have already been described. We have come far from the number known to Linnaeus, i.e. about fifteen. However, what is still more remarkable is that the number of new species has doubled in less than twenty years just as if these peculiar and beautiful plants had been patiently waiting in their unexplored solitude to be pushed out into the light of science.

The traveler-botanist who becomes attracted to this task does not fail to report on fine and numerous new species. The field is open and the worker does not lack for work, nor does the work lack for workers.

—Edouard Francois Andre. 1889
Bromeliaceae Andreanae.

Chapter 1
Introduction to Tillandsia

Introduction

These rather bizarre looking plants come from the family of flora, Bromeliaceae (bro-meh-lee-AH-say-eye), more commonly referred to as BROMELIADS. The bromeliad family includes a wide range of plants, such as the pineapple, *Ananus comosus*, and the famous "Spanish Moss," *Tillandsia usneoides* Linnaeus.* A plant family name is used to describe a broad variety of related plants. Cactus (cactaceae) and Orchid (orchidaceae) are the common names for plant families with which you are probably familiar. More closely related plants in these families are separated into GENERA. Examples of two orchid genera would be *Cattleya* and *Cymbidium*. Each individual type of plant is then given a species name that separates it from all other plants.

The bromeliad genus, *Tillandsia*, is named for a Finnish physician, Elias Tillands (d. 1693). It is an epiphytic (able to grow without soil) genus comprising over 400 recognized species, largest in the family. Even today, more species are being discovered in the remote corners of Latin America as dedicated enthusiasts seek them out. Some *Tillandsia* species are ubiquitous, growing over thousands of square miles. Many number in the millions in their native habitats. There are also a few exotic species that are confined to a single valley or mountain range. Because of their specialized growth requirements, they are difficult to cultivate and generally are not available.

With the exception of some of the southern parts of the United States, all tillandsias come from Latin America. Although the majority of species flourish in MESIC (wet) habitats, most popular species of *Tillandsia* proliferate in XERIC (subject to drought and/or aridity) environments. They are often found growing exposed to sunlight, perched singly or in clumps on various types of cacti, thorny scrub bushes, or trees, rock outcroppings, cliffs, or even directly on the gravelly or sandy ground.

Tillandsias are true air plants, unlike the popular "air fern," which is actually a green-dyed seaweed from Great Britain. Xerophytic tillandsias are able to exploit what we would consider to be harsh habitats—that is, droughty and hot—by the development of very interesting mechanisms that are discussed in some detail in the sections on Foliar Trichomes and Trichome Variations. Suffice it to say here that among the tillandsioid modifications are: the development of a thick layer of water storage tissue, which increases the number and shield complexity of the absorbing trichome (white fuzz on the leaves), a decrease in leaf number, a decrease in root development, a decrease in overall plant size and impoundment capacity, a reduction in the number of flowers and fruits per shoot (plant), and an increasing tendency for self-fertilization.

Xeric-growing tillandsias have evolved to be stress tolerant because extreme epiphytism is possible only under certain very rigid criteria. Because of these adaptations, tillandsias are poorly designed to grow large quickly.

Tillandsias are tremendously adaptable, tolerating a wider range of conditions than most other plants. After being placed in a good location and having made the adjustment to their new surroundings, they are capable of surviving for long periods of time without attention. Take care of them, but do not be a slave to them. They will manage just fine!

Physical Characteristics

Tillandsias vary greatly in size, shape, and texture. A strand of *Tillandsia usneoides* is very small indeed, with 3 cm. or 4 cm. constituting a healthy and viable plant. Yet a mesic species, e.g. *Tillandsia grandis* Schlechtendal, sometimes grows to be five meters tall when in bloom! Mesic tillandsias such as *T. flabellata* Baker originate in rainforests and are soft-leaved, while xerophytes like *T. jucunda* Castellanos have stiff leaves. *Tillandsia ponderosa* L. B. Smith and *T. yunkeri* L. B. Smith are rosette-shaped like the well-known *Aechmea fasciata* Baker. These plants have leaves emanating from a single point in the center of the plant, forming a well in which the plant impounds water among the leaf axils. Species such as *T. caput-medusae, T. bulbosa and T. butzii* appear to be from another galaxy, let alone another planet. One beautiful species, *T. plumosa,* appears to be a snow-covered sea urchin.

This fuzzy-leaf covering consists of foliar hairs called TRICHOMES or PELTATE SCALES. Although the trichomes are present in many forms of plant life, they are very highly developed in xeric species

*oos-nay-oh-EE-des.

of *Tillandsia*. When it rains or there is dew in the air, these scales rapidly transfer water from the leaf surface to elongated storage cells inside it. As discussed in the section, Foliar Trichomes, the scales aid the plants in a number of other important functions. In some species that receive moisture at frequent intervals, the trichome wings are vertical to the leaf surface in order to facilitate epidermal drying. In others, the wings parallel the epidermis to help insulate the leaf from extreme heat. Another valuable function performed by the trichomes is to reflect the excess sunlight that might damage the leaf.

Tillandsias that live under mesic conditions have smaller, widely dispersed trichomes. Often, they cannot be seen with the naked eye. *Tillandsia tectorum* has perhaps the largest trichomes, and these are easily visible. The reflection of light off these trichomes is what gives the plant its bright white countenance.

One of the most remarkable characteristics of this genus is the incredible diversity of species within it. Probably no other genus, and few families, are so vastly different in MORPHOLOGY (appearance). Why are so many species so different from each other? This intriguing question is answered in more detail in the section on Evolution.

Blooms and Reproductions

Beautiful flowers are another remarkable feature of these plants. The intensity and richness of the purples, reds, pinks, yellows, and greens is truly something to behold! When you notice the plant beginning to send out its inflorescence, more light may intensify the colors. The blooming cycle for some rapidly growing species may be only two or three weeks. For most, it is a month or two, and for some of the most xeric, slow-growing species, it may last as long as a year. A number of species, such as *Tillandsia xiphioides, T. duratii, T. purpurea*, and *T. straminea*, have wonderfully fragrant, long lasting blooms. The flowering season for tillandsias usually begins in late fall or early winter and runs through early spring. In the United States, however, there always seems to be something blooming in a fair sized collection.

During or after blooming, tillandsias produce anywhere from one to a dozen baby plants called offsets, offshoots, or pups. Many produce pups from the base of the plant, while others grow offshoots from the base of the inflorescence. The former offsets are termed BASITONAL while the later are ACROTONAL. Some produce numbers of pups from among the leaf axils along the stem in CAULESCENT (stem-growing) species. Some species, such as *Tillandsia latifolia*, produce pups from the inflorescence itself, and a few, such as *T. tricolor*, grow pups at the end of short runners called STOLONS.

Pups gradually grow to adult size in one to four years, while the parent plant gradually dies over the space of a generation or two. In this way, along with seed production, the species is perpetuated, increasing its numbers through the generations. Although the parent plant eventually dies, in most species, it keeps up its appearance until the pups are fully grown. Since each parent usually produces more than a single offset, the plant will increase in numbers over the years forming a beautiful, multiblooming clump.

A few *Tillandsia* species produce very large bloom spikes with thousands of seeds. The total energy of the plants is devoted to this effort, so they produce no offsets. These MONOCARPIC species are some of the larger, mesic TAXA (taxonomic unit, e.g. genus or species), such as *T. prodigiosa* Baker.

Many people find their plant will not bloom if kept in the house. This is usually because a person's indoor environment does not duplicate closely enough the natural environment. Most often, the plant does not receive enough light and/or moisture to sustain a concerted blooming effort. In other words, the plant is not strong enough. Or maybe the length of time the plant receives sufficient light is not long enough to trigger the blooming cycle.

If your plant is mature, keep it outside and eventually it will bloom if given the right conditions of sufficient light, air circulation, temperature, and moisture. Once the inflorescence has developed, bring the plant indoors to enjoy it if you desire. Interestingly, a number of experienced collectors sometimes do not want a particular *Tillandsia* to bloom. For once begun, the cycle is irreversible, and that particular plant will not grow any larger. It will reproduce, but the shoot will have seen its day, and if the specimen has taken a number of years to grow to magnificent, unheard-of proportions, one can understand someone's desire in wishing not to lose an old friend.

For those who really want to see their plant(s) bloom, it can be done artificially. The plant should be strong and healthy before flowering is induced. There are two basic methods, both of which work well. One is to buy a product called Florel at your local nursery. A little solution is mixed with water and sprayed onto the plant. The other way is to place the plant in a plastic bag with a piece of ripe apple for a couple days. The ethylene gas from the apple will do the trick in 6-14 weeks, depending on the species. Many people get more satisfaction if the plant is allowed to complete its life cycle naturally. In addition, the bloom tends to be longer lasting.

Tillandsia pups can be left attached to the parent plant, eventually forming an attractive clump, or they can be separated and attached to something else to grow singly. If you choose to separate the plants, wait until the pups are half the parent size before removing them. Use a sharp knife or pair of clippers and cut them off cleanly at the point of attachment. Keep the wound clean and let it dry for a day or two. If you want the *Tillandsia* to root to the surface of the mounting material, it should be rough or filled with crevices to give the roots something to grab hold of. The plant can be glued, wired, pinned, or tied with fishing line or strips of nylon stocking material.

As previously mentioned, tillandsias reproduce by seed and offset. In their native habitats, they have adapted to conditions of often brisk air movement and intense sunlight. If tillandsias produced seeds like most other plants, most of them would fall to the ground, and the species gradually would become extinct. But the plants have evolved a fuzzy, cottony attachment on the seed coat called a COMA. This parachute-like apparatus allows the seed to float on air currents for long distances, thereby allowing them to maintain their habitats high in trees, on cliffs, etc. It also allows them to rapidly colonize new habitat sites, and establishing these FOUNDER POPULATIONS is one of the important factors involved in the great MORPHOLOGICAL (physical) diversity within the genus. The coma hair, like the trichome, indicates a high degree of specialization.*

Outdoor Care

One of the most compelling aspects of tillandsias is the ease with which they are maintained in an outdoor patio or under a tree. In coastal or humid environs, many will flourish in full sun.

As a general rule of thumb, the greener, softer-leaved plants require less sun and more moisture, and they tend to thrive in cool environments. The grayer, stiffer-leaved species prefer more light and less moisture and are usually from warmer areas. One should acquire tillandsias that will take advantage of the locations in which they are to be grown.

Most xeric species come from climates that are often humid so they do not TRANSPIRE (lose water) as quickly as they would in a dry environment. The rate at which a plant will lose its internal water supply (the soft tissues of a bromeliad are 80-90% water) is a function of temperature and humidity. The higher the temperature and the lower the humidity, the faster the plant will lose water.

During periods of low humidity, intense sunlight can burn the leaves of even the most xerophytic *Tillandsia* unless the plant has been gradually acclimated to long periods of direct light. In much of the eastern, midwestern, and southern United States where it is quite humid, the problem of too much light is not as critical as it is in other, drier areas.

Fig. 1.1 A blooming clump of *Tillandsia chlorophylla* L.B. Smith. It is a mesic species that was put in sunlight to photograph.

The ability of tillandsias to survive long periods of drought (owner neglect?) qualifies them to be among the world's hardiest plants—especially when combined with their ability to withstand near-freezing temperatures, extreme heat, high wind, etc. In combination with the fact that there is no soil to contend with, they also qualify as one of the world's most carefree plants for growers—if one has a basic understanding of how the plants function.

Tillandsias love fresh air breezes (particularly in hot, humid conditions). Of course, with more exposure to actual wind, they dehydrate and should receive more water. Temperature is not much of a factor unless it either approaches freezing or gets quite hot. The mesic species are more susceptible to extremes of cold and heat. If the environment is hot and dry, overwatering the plants is almost impossible. In fact, under these conditions, the more tillandsias are watered and fertilized (assuming they are receiving strong light), the faster they will grow.

At Rainforest Flora, Inc., we water the plants, except *Tillandsia tectorum,* from one to three times a day in the hot, dry summer, and the plants grow rapidly. Xeric tillandsias cannot be overwatered unless the epidermal surfaces remain wet for longer than a couple days. In this case, the leaves will not be able to exchange vital gases and may suffocate, as a human who drowns from lack of oxygen. This is explained in more detail in the section on Water and Moisture Aspects.

If growing conditions include low humidity, tillandsias can be grouped close to other soil-bound plants. The moisture of the overall group of plants will help humidify the surrounding air, creating a

**See* section on *Tillandisia* Seed, pg. 202.

livable microclimate. Another way to accomplish this is to fill a large, flat pan with gravel and water and place the mountings on it.

As a rule, tillandsias are not bothered by insect pests. Scale and mealy bugs are the most common, and they are easily eradicated. The plants can be dipped in a pesticide solution of malathion or cygon (preferable) and a tad of liquid dishwater detergent (the detergent acts as a surfactant that allows the water to spread across the leaf surface instead of beading on it). Usually a tablespoon per gallon of pesticide will do the trick. The plants are then soaked for a few minutes and subsequently allowed to dry; they should not be rinsed.

If a *Tillandsia* should begin to rot—that is, the center leaves fall out—it should not be discarded unless there is nothing left alive. This may be checked by taking the plant and pulling gently on the centermost leaves. If they come out easily and the tips are black, the process should be continued in circular fashion from the center outward. If a point is reached where resistance is felt and there is a substantial portion of the plant left, it can be saved.

The specimen should be dried completely and then treated as any other *Tillandsia* because it will often produce numerous pups. However, if the pulled leaf tips are white, the plant was healthy and the leaves were pulled too hard.

Sometimes part of a *Tillandsia* base may become soft and mushy like a bruised apple. A clean knife is used to excise the bruise. All of the brown area must be removed or the rot will spread. The base should be kept dry until it callouses over, and then the plant should be treated as before. If the rot has not spread too far, the plant will soon root and produce new growth.

Indoor Care

Tillandsias kept in the house longer than a month need to be watched closely until they establish themselves in an environment. They love fresh air, good light, and humidity—conditions often absent in the home. However, since tillandsias possess the ability to adapt to a wide range of climatic conditions, they often will grow (or at least not decline) indoors if they are given as much of their natural surroundings as possible.

Tillandsias kept in the house should receive plenty of strong light from a nearby window (preferably facing east, west, or south). If this is not practical, broad spectrum fluorescent lights are available that provide 92% of actual sunlight when placed 15-30 cm. above the plants.

Watering is critical indoors since there is usually a lack of humidity, especially in homes or offices with air conditioning and/or central heating. A successful way to water the plants is to totally submerge them every week or two in a sink, bathtub, or bucket in room temperature water containing a small amount of fertilizer. They should remain submerged overnight, or at least for a few hours.

Soaking the plants in this manner will allow access to enough water to overcome any deficit. Although the plants cannot respire while under water, it will not be critical unless extended for more than a day or so. At that point the plants may suffocate from a lack of carbon dioxide and oxygen respiration.

Misting also helps in dry conditions, but this should be done in addition to thorough soaking, not in lieu of it. In dry conditions, water misted onto a plant often evaporates before the plant has had the opportunity to absorb it. Even if it does absorb water misted on the epidermis, it may not be enough to overcome a water deficit. The plants will gradually dehydrate and may eventually die from a lack of moisture.

In a light airy house, most tillandsias will adapt. In a stuffy, closed-in environment, they may not. Success varies from person to person, household to household, species to species, even plant to plant. If the plants thrive, terrific. If they begin to look poorly, they should be moved outdoors.

One foolproof method for enjoying the plants indoors is to rotate them every month with plants that are growing in better, outdoor conditions or under wide spectrum fluorescent lights.

Fig. 1.2 With plenty of bright light and a biweekly, overnight soaking, many tillandsias thrive in the indoor environment.

Fertilization and Water Quality

Xeric tillandsias are supreme OLIGOTROPHS—nutrient scavengers. In nature, they receive little in the way of nutrients; however, they make maximum use of those nutrients that are available. This does not mean a generous fertilization program is useless. To the contrary, a consistent fertilization program during the warm months will produce much larger, more robust plants when combined with strong light and frequent, thorough watering.

The most important nutrient in terms of plant vigor and foliar growth is nitrogen. Most commercial fertilizers contain a high percentage of urea-based nitrogen. However, the nitrogen must be broken down by bacteria in soil before it becomes available to the plant. This is fine for potted plants, but not for epiphytes that have no soil-bound roots. Ammoniacal or nitrate nitrogen is immediately usable, but many fertilizers have little in their mix.

Potassium and phosphorus are

critical to plant health. Among other things, they enhance flowering and seed setting, help the plant resist disease, promote root growth, and strengthen cell walls.

A highly recommended fertilizer is EPIPHYTES DELIGHT. It contains the two useful forms of nitrogen and little that is urea-based. Besides the necessary phosphorus and potash, this well balanced source of nutrients also contains eight essential chelated trace elements so the plants can grow to their full potential.

Epiphyte's Delight is designed to provide a plant food ratio of 2:1:3, which is believed to be the optimum ratio for balanced growth. The plants will grow faster, larger, and produce more vibrant, impressive blooms and more offsets when fertilized in combination with a program of bright light and the supply of sufficient water.

The plants should be fertilized on a consistent schedule. A very dilute amount every time the plants are watered would be ideal. But fertilization every week or two with a stronger solution is fine. If fertilization is performed on a biweekly basis, up to a ½ tablespoon per gallon ratio may be used. The plants are misted or submerged in the solution of fertilizer and water.

Water does make a difference in the appearance of the plants. Many city dwellers have alkaline water in their public system. Over time this can cause a white salt and mineral buildup on the leaves that is unsightly. Bromeliads in general, and tillandsias in particular, prefer slightly acid (pH 6.0) water. A pH of 7.0 is neutral. City water often has a pH of 8.0 or more, which is quite alkaline.

Epiphyte's Delight contains sufficient quantities of an acid-forming compound (phosphoric acid) to counteract most alkaline water conditions. If your water is alkaline, mix ½ tablespoon of Epiphyte's Delight in one gallon of water and let it stand overnight. This will give the fertilizer time to precipitate much of the dissolved minerals and salts in the water. A fine dustlike material will collect in the bottom of the container. Do not apply this to the tillandsias, but use it on your garden plants instead.

For plants that are submerged in water and not fertilized as part of their regular care, a little vinegar, sulfuric, or phosphoric acid added to the water will drop the pH. A pH test kit can be purchased from a swimming pool supply store or an aquarium shop.

Mounting Tillandsias

Tillandsias are most often marketed on pieces of sandblasted or tumbled grapewood, which is relatively light and easy to work with. Pieces of driftwood, bark, manzanita, etc., also serve the purpose of providing an attractive background and mounting for tillandsias.

Mounting most tillandsias is simple. Smaller species may be glued directly onto the surface of the wood. A natural depression in an appropriate mounting spot may be used for the root base. If one wishes, a drill can sculpt a small hole or depression to better anchor the plant. However, care should be taken to avoid attaching the plant deep into a hole that will cover up much of the base.

There are advantages and disadvantages to every type of glue, but generally, any adhesive will accomplish the task. We use hot glue because it dries in a minute. When hot glue is used, care must be taken with small species such as *Tillandsia argentea* and *T. ionantha*. The glue comes out of the gun at 380 °F. If this heat reaches the meristematic (growing tip) tissue, the plant may die. If the plant has a good root base on which to put the glue, the process should not be a problem if one is careful. One tip is to wait 10-30 seconds before attaching the plant. This gives the glue on the mounting medium a chance to cool a little.

Many species are AGEOTROPIC, that is, they grow irrespective of gravity and can therefore be mounted in any position—even upside down. The plants can be creatively mounted on hanging pieces of wood, using all its dimensions. Some species, such as *Tillandsia incarnata* are CAULESCENT (grow along a stem) and can be mounted in a hanging position. The apex of the plant will grow upward toward the light source. Species like *T. magnusiana* and *T. atroviridipetala* should be mounted in a horizontal position so water cannot accumulate among the leaf axils, which could cause the plant to rot.

Larger species, such as *Tillandsia xerographica, T. fasciculata*, and *T. rodrigueziana* Mez can be glued when mounted in a sitting position. But when the plants are meant to be hung, they are normally too large to be mounted effectively with glue. Wire or fishing line is often used. The wire or line should be made as unobtrusive as possible. This is accomplished by threading it behind the outer leaf sheaths (bases). Then the wire or line can be threaded through the wood, the material is twisted or tied tightly at the back, and, if not performed too low on the plant, the specimen should be stationary.

A word of caution! Plants that are stuck into holes often rot when the base becomes wet and does not have an opportunity to dry. This mistake occurs often when people stick the plants down into the opening of a seashell. Another problem is sometimes encountered when a giant swath of sphagnum moss is wrapped around the plant base. This is usually done to make the whole mounting look more attractive. But again, it blocks air circulation,, and will often hasten the demise of a plant if kept moist. *Tillandsia punctulata* and *T. cyanea* are about the only widely available tillandsias that prefer this moist arrangement, and that is because they are mesic species.

Sphagnum moss, or any other such medium, is normally used as a cosmetic device to hide the glue or wire used to mount the plant. As long as only a small amount is used, this is fine. The knowledgeable grower knows the purpose of the moss and will not keep it excessively wet.

Many mesic species of *Tillandsia* are terrestrial and prefer to be potted in soil. A good soil mix is fast-draining, porous, and slightly acidic.

Decorating and Landscaping with Tillandsias

Tillandsias can be used with great effect in decorating or landscaping. They are so varied in morphology and so versatile as to care conditions that virtually the only real limits involved are one's imagination.

Tillandsias can be attached to tree branches and grown as in nature. They can be suspended in midair on fishing line or wire. They can be mounted on tree stumps, rocks, patio posts, or anything else one has lying around.

Certainly tillandsias can be grouped together and grown as a collection—it is the way most people grow them. But they can also be used to create great depth in an overall garden situation. Appropriately placed, tillandsias can show off areas not available to other plants.

When attaching plants to a tree or shrub, certain facets of plant care should be kept in mind. One should remember that most *Tillandsia* species will grow larger and faster with higher light levels. Therefore, most of the species should not be attached deep in a tree canopy where they will be afforded little light—at least if they are to be watered frequently.

The plants can be mounted on tree branches at all angles, as this is the way they grow naturally. One never sees the branch of a tree with all the tillandsias lined up like so many sentinels along the top of a branch. Experimentation and imagination are the keys. Some species can be grown well upside down. Most will thrive if mounted horizontally. Of course tank-type species should be attached vertically so the tanks can fulfill their water-impounding purpose.

By grouping three or four plants of the same species together, the grower can save a year or two in the time it takes to develop a clump. This is often a good way to go with smaller and/or slow-growing species.

Fig. 1.3 The author's wife, Elizabeth, modeling the latest in decorative *Tillandsia* wigs.

Species such as *Tillandsia incarnata, T. sphaerocephala, T. bergeri*, and *T. tectorum*, along with many others, grow saxicolously on rocks or cliff faces. This natural way of growing the plants can be used with great success in one's garden, and virtually all the xerophytic species can be grown in this manner.

In the patio, a wall is often the logical place to start decorating with the plants. If the patio is well lit, it provides the perfect environment. One or two large specimens strategically placed will normally set the pattern for how and where future plants will be positioned. Patio posts make ideal settings because they are usually open, airy, and receive bright light. If the posts are wooden, bareroot plants eventually will root to the post and, after a few years, form an attractive clump.

Tillandsias can be attractively mounted on pieces of wood, stones, sea shells, bric-a-brac, or prized ceramic pieces. One enterprising UCLA student marketed tillandsias mounted on old tennis shoes that had been preserved with a special resin coating! It makes no difference what the mounting material is, as long as it is not something toxic. Much more important are the conditions in which the plant is cultivated.

Tillandsias as Animal Hosts in Nature

As a rule, tank-type bromeliads, including mesic tillandsias, play host to a great variety of animals. Entire ecosystems of worms, arthropods, reptiles, amphibians, and insects live out their entire lives in just one or a few plants. For some creatures, water impounding bromeliads provide the only source of water for part of the year. Most of these creatures are not considered pests.

The pests most commonly found on xeric tillandsias are the armored scales (family *Diaspidae*) and the soft scales (family *Coccidae*). These insects are not often encountered, but when they are, an entire population of plants may be infested. Another insect pest is the mealybug (family *Pseudococcidae*). Mealybugs most often flourish at the base of a plant among the leaf axils.

Pests on Tillandsias in Cultivation

As in the wild, the two most commonly encountered pests in gardens, at least in Southern California, are mealybugs and scales. Both may be eliminated by submerging the plant for a few minutes in a pesticide bath. Although malathion is often prescribed, some consider it to be too weak to always achieve 100% per control. Cygon (dimethoate) does provide control, and it is not considered a strong or dangerous pesticide. Once mixed, the solution should always be used within 24 hours. After this time the water will have broken down the active ingredients into harmless byproducts. Both of these pests are a food source for the ladybug (family *Coccinellidae*), which can provide a certain amount of control.

Female scales are mobile only as newly hatched nymphs, or "crawlers." The crawlers of both sexes disperse and find a suitable spot on which to settle. Scales then grow and molt, producing a second nymphal stage in which the sexes are only slightly differentiated. After the next molt, males begin to develop wings and copulatory organs. Interestingly, adult males lack the stylet (stiff, tubelike mouth part used for feeding) of true bugs, so they cannot feed. Female adult scales develop a waxy shell. Then they insert their mouth parts into the leaf and feed.

As can be seen in the accompanying photo, mealybugs are fluffy and white. They are mobile and often introduced to the plant by ants who use their sugary exudate as a food source. Mealybugs are covered with a white or yellowish, powdery wax. Waxy filaments often are found around the body. They develop in colonies and excrete large amounts of honeydew. Mealybug eggs, which are yellowish or orange, are laid in the masses of white, cottony sacs. The ground mealybug, *Rhizoecus falcifer*, is smaller, with a body covered evenly with white wax. Living in the soil exclusively, it attacks the roots of terrestrial plants.

Both scales and mealybugs belong to the order of insect, Homoptera. Homopterous insects are PHYTOPHAGUS (plant eating) and puncture the sieve tubes with their PROBOSCIS. They suck the plant juices from the vascular tissues. The internal turgor pressure on the fluid pushes it out and saves the insect the work of having to suck. The juice of the tillandsias, however, contains more carbohydrate than the insects need. Scales and mealybugs cope with this surplus in two ways. First, the sugar can be passed out of the body as liquid excrement, dropping onto the leaves or ground, or it may be sucked from the CORNICLES (growths on the backs of the insects) by hungry ants. The second method used by scales and mealybugs is to convert the excess carbohydrates into wax within the body; this is then secreted by glands on the body surface as threads, or as an overall covering.

Snails are rarely a problem,

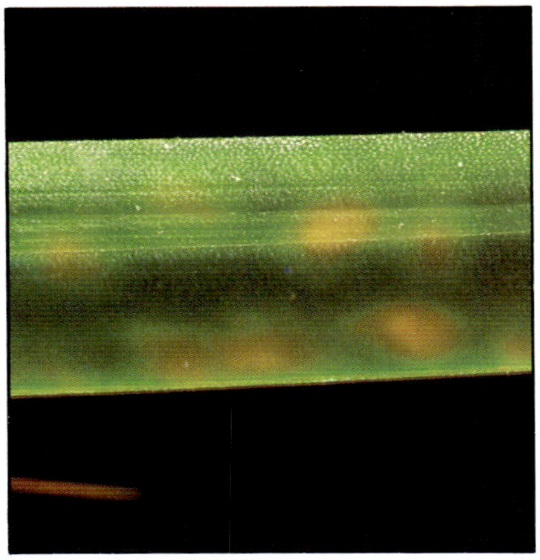

Fig. 1.4 Adaxial epidermis of *Tillandsia utriculata* showing damage caused by the pineapple scale, *Diaspis bromeliae* Kerner.

Fig. 1.5 Closeup of abaxial surface of same leaf showing actual scale infestation.

Fig. 1.6 Infestation of male and female scale. Males are white and linear. Females are round and covers have been removed, exposing the yellowish body beneath (photo Parrella).

with the exception of mesic species such as *Tillandsia fraseri* Baker. In a garden setting, however, they do have a tendency to take advantage of the protection offered by thorny specimens of *Aechmea, Neoregelia, Bromelia,* etc. In California the common garden snail, *Helix aspersa*, is sometimes a real problem, and clusters of snails can be encountered buried among the leaf axils. To a lesser degree, this applies to slugs (family *Arionidae)* as well.

Reference Material:

Drummond, W.C. 1954. Pests on Bromeliads. Los Angeles, California. The Bromeliad Society Bulletin. Vol. IV, pgs. 3-5.

Grzimek. 1975. Grzimek's Animal Life Encyclopedia. New York, N.Y. Von Nostrand Reinhold Co. Vol. II. Insects. Vol. III Mollusks.

Fig. 1.7 A severe infestation of mealybugs preventing root development at the base of *Tillandsia fasciculata*. This specimen had been cultivated in soil.

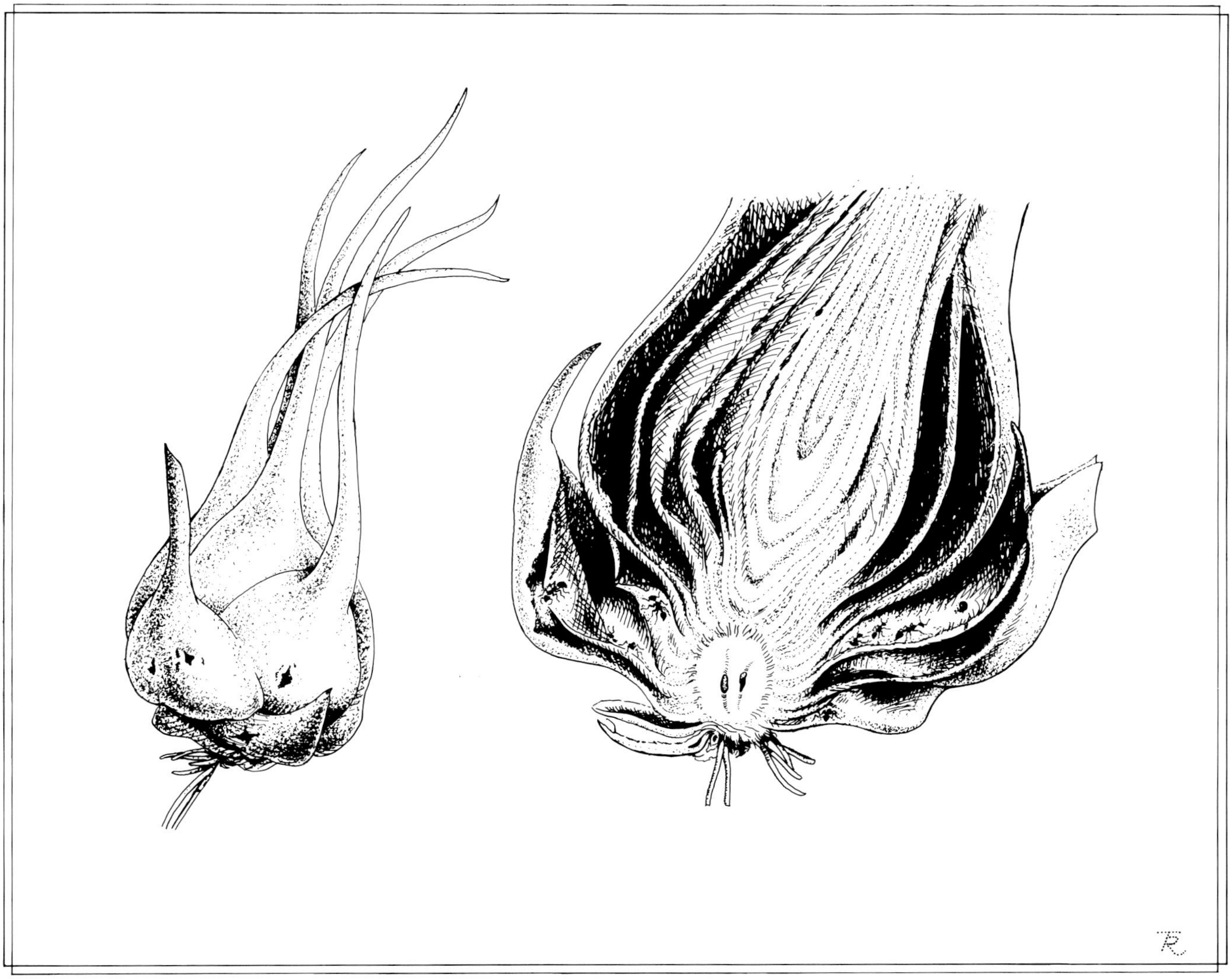

Fig. 1.8 Drawing of *Tillandsia seleriana* showing artist's conception of ants in plant interior.

Myrmecophytism in *Tillandsia*

A number of species, especially in subgenus *Tillandsia*, have enlarged bases. The inflated, convex sheaths form hollow chambers within the base. Plants exhibiting this feature are termed pseudobulbs, or "false bulbs." Pseudobulbous species often form an interesting symbiosis with ants, which qualifies them as MYRMECOPHYTES, or "ant plants." An entire colony may live inside a plant, gaining protection from predators and the elements. The plant, in return, receives nutritional benefit from the FRASS (excrement), cast skins, and other refuse of the ants.

Along the Rio Dulce in Guatemala, I once encountered a large *Tillandsia streptophylla* that had colonies of two different ant species living in it. The ants of one colony were significantly larger and a different color than those of the other. The two colonies lived in separate parts of the plant.

In order to gain access to the plants, the ants chew holes through the sheaths. Many collectors feel that these holes add an interesting bit of lore to the plant, and purposefully select those that have this extra piece of ant handiwork.

My experience has been that ants in Southern California have been known to inhabit tillandsias (usually not pseudobulbous species), but the degree to which they do so is not normally significant. They do not chew holes in the leaves, although they may cultivate mealybugs or scale.

Chapter 2
The Species

Tillandsia aeranthos

L.B. Smith; 1943

(eye-RAHN-thos)

subgenus: *Anoplophytum*

***Aeranthos* is a latinized combination of the two Greek nouns *aer* and *anthos*, which mean "air" and "flower" respectively. Therefore, *aeranthos* means "air blooming." The botanist Loiseleur described this species as *Pourretia aeranthos* in 1821. In 1943, it was included in *Tillandsia*.**

Tillandsia aeranthos grows epiphytically from near sea level to several hundred meters in Argentina, Uruguay, Brazil, and Paraguay. It has a preference for mature *tipa* trees *(Tipuana tipu)* and tends to grow singly or in small clusters of two to a dozen plants.

A caulescent (growing along a stem) species, *Tillandsia aeranthos* has an average dimension of 11-13 centimeters in diameter and height. The leaves are spreading to erect, densely polystichous (leaves emanating in all directions from the stem), and covered with appressed (wings of trichomes laying flat, parallel to epidermis), cinereous (ash-colored) trichomes. The leaf blades are narrowly

C.

triangular and attenuate (long, tapering to a point).

This species is virtually indistinguishable from *Tillandsia bergeri* in the non-blooming state.* Both *T. aeranthos* and *T. bergeri* can be used to indicate the state of species hydration or dehydration (desiccation) in one's collection. The succulent leaf blades are only slightly channeled when hydrated, but they are deeply involute (leaf edges curled up) and thinner when dehydrated.**

The slender scape exserts the simple (single) bloom spike beyond the leaves. The floral bracts are ovate (egg-shaped and attached at the widest end) and ruby red. The flared petals of the polystichous flowers are often an indigo so deep as to appear almost black. *Tillandsia aeranthos* blooms annually, and the entire flowering cycle lasts close to a month. *Tillandsia aeranthos* produces roots quickly when well cultivated, and readily attaches itself to the mounting substrate.

The colorful bloom of *Tillandsia aeranthos*, as well as those of other species, has won admiration among the Argentines, who refer to them as *clavelles del aire*, or "carnations of the sky." In addition, *T. aeranthos* is probably the only species to be immortalized in song. The tango *Clavel del Aire* was scored and lyricized by Juan de Dios Filiberto and F. Silva Valdez in 1935.***

This species does not develop as many offsets over its life cycle as does *Tillandsia bergeri*. However, it usually produces three to six after blooming and, after a few years in cultivation, it too develops into a large clump.

**See* description of *Tillandsia bergeri* for more discussion.

***See* section on Water and Moisture Aspects in Growing Tillandsias, pg. 187.

****"El Inolvidable Filiberto"*; Copyright 1934/35, EMI-ODEON SAIC (Argentina).

A.

B.

A. With good growing conditions, a large clump of *Tillandsia aeranthos* may develop in six years.

Tillandsia albertiana

Vervoorst; 1969
(all-bear-tee-AH-na)
subgenus: *Anoplophytum*

***Tillandsia albertiana* was named for the Argentine botanist Alberto Castellanos, by an admiring student who discovered the species.**

Tillandsia albertiana is restricted to a small area of the Andean foothills in northwestern Argentina. This saxicole, or rock dweller, grows in dense clumps on cliffs overlooking mountain streams. The terrain is rugged and the weather unpredictable in this part of the country.

Recently described, *Tillandsia albertiana* is truly a jewel. The bloom is enormous in relation to plant size and vibrant cherry, which is rare for a *Tillandsia. Tillandsia edithae, T. schiedeana* v. *glabrior* Gardner, *T. andreana* E. Morren, and *T. funkiana* Baker are the only others with red flowers. The solitary bloom of *T. albertiana* usually lasts for days.

Tillandsia albertiana is a small plant, with an adult usually averaging 5-8 cm. in height and width. It generally produces a large, single root. The plant offsets prolifically so if the grower groups three or four of them they may develop into a sizeable clump in two or three years.

The shape of this caulescent species is a work of Nature's art. The leaves are distichous (growing in mirror fashion opposite one another along the stem) and sublinear (long and narrow, the

D.

sides almost parallel). This gives the plant a two-dimensional aspect. The leaves are spreading and make a graceful arch, stacked evenly one above the other in almost perfect proportion. They are covered with appressed, cinereous trichomes. An interesting feature, shared with *Tillandsia argentina*, is that the abaxial (outer) blade surfaces are often slightly wrinkled.

My partner, Jerry Robinson, and I made the locating of this plant a top priority when on a trip to Argentina in 1982. We hired a pickup truck and driver and went off to find it. As we ascended the Andes, the road became eroded in places by swollen streams and troughs of mud. Determined to continue, we were slowly able to forge ahead until at last the road turned into a sandy cow trail and went steeply downhill for a long distance. We were 20 kilometers short of our goal and we were stuck.

As our contingent happened to be on private property, a 75,000-acre *estancia* to be exact, we decided to ask the administrator of this ranch if we could rent horses and perhaps a *gaucho* to help us the rest of the way. Well, Dame Fortune certainly smiled on us that day! The administrator—one Guillermo Smith, son of Scottish immigrants—welcomed us with open arms. He sent the chauffer packing and invited us to stay with him for a few days. Guillermo spoke a perfect Scot's English and was an entertaining, gracious host.

The *estancia* was paradise! The experience of heading back into the pristine, uninhabited mountains, so abundant with wild animals, birds, trees, flowers, and grasses, was a rare thrill. What a great reward for all the times when exploring and collecting has been frustrating, exhausting, and sometimes dangerous work.

Like *Tillandsia argentina, T. albertiana* grows in areas somewhat shaded from all day sun. Moderate patio conditions provide the perfect environment for this delightful little plant.

A.

B.

C.

A. Plant enlarged approximately two times.

B. *Tillandsia albertiana* can be seen growing in large clumps above the heads of author and *gaucho* (photo Robinson).

C. The two corolla shapes of *T. albertiana*; the form on the left is much more rare.

Tillandsia argentea

Grisebach; 1866

(are-GEN-tay-a)

subgenus: *Tillandsia*

***Argenteus* is an adjective that is derived from the word for silver, *argentum*. Argenteus means "silvery," a reference to the leaf color.**

Tillandsia argentea grows epiphytically in dry, woodland habitats in Mexico, Guatemala, Cuba, and Jamaica. This species has a bulbous base and tends to grow singly or in clumps of just a few plants.

Tillandsia argentea is similar in appearance to *T. filifolia* in the vegetative state; both species have bulbous bases and filiform (thread-like) blades. An adult specimen of *T. argentea* is normally 10-12 cm. in diameter, and the blades and sheaths are silvery blue-green because of the dense covering of cinereous trichomes. The leaves of *T. filifolia*, on the other hand, have fern green blades which is the result of more widely dispersed scales that have smaller wings. The sheaths are grape violet to castaneous (dark brown). *Tillandsia filifolia* normally grows larger than *T. argentea*.

C.

The inflorescence is glabrous (smooth, not hairy), simple, the blooms few in number. The scape and floral bracts are cherry and the corollas (flowers) are grape violet. The morphology is varied. The Mexican form often becomes larger and more robust than the finer-leaved, more delicate appearing Guatemalan form.

Tillandsia argentea grows well under moderate *Tillandsia* conditions. The plant should receive bright light and sufficient air movement; the amount of water will depend on the humidity and temperature of the grower's microhabitat.

With proper cultivation, the plant grows to be beautifully symmetrical; soft and fragile in appearance yet hardy and fast growing. It usually does not grow well as a year-round houseplant. As with other filiform-leaved species, such as *T. filifolia* and *T. juncea, T. argentea* prefers an airy, outdoor environment.

As much as any other species, *Tillandsia argentea* shows off the complementary aspects of its different plant parts. This is one of the most exquisite and refined looking species of *Tillandsia*.

A.

B.

A. Guatemalan form of *Tillandsia argentea* on the left.

B. Mexican form on the right.

Tillandsia argentina

C.H. Wright; 1907

(are-gen-TEE-nah)

subgenus: *Anoplophytum*

***Tillandsia argentina** was named for the country in which it is found.**

Tillandsia argentina grows as an epiphyte (grows on trees or shrubs) or saxicole (grows on rocks) in northwestern Argentina at elevations of 400-1,600 meters. It is usually found in large clumps with the individual specimens growing together so densely that it is difficult to tell where one plant ends and another begins. *Tillandsia argentina* often inhabits evergreen trees along with *T. tenuifolia*.

Tillandsia argentina is a small, caulescent species, normally 5-7 cm. in diameter. Interestingly, the abaxial leaf blade surfaces are quite wrinkled—more than the blades of *T. albertiana*. This aspect of its morphology does not seem to be related to water stress because the leaf does not change whether hydrated or dehydrated.

The blades are similar to those of *Tillandsia albertiana*—long, slender, and stiff. The two plants are easy to tell apart because the leaves of *T. argentina* are polystichous while those of *T. albertiana* are distichous. The

D.

leaves of *T. argentina* are lepidote (covered with fuzzy trichomes) and olive green while those of *T. albertiana* are more glabrous (smooth, not fuzzy) and forest green. The leaves of *T. argentina* are erect, slightly spreading, and covered with large, appressed (wings of trichomes laying flat, parallel to epidermis), cinereous (ash-colored) trichomes that under a magnifying glass have an interesting honeycomb appearance.

The inflorescence is simple (unbranched), complanate (flattened), and lanceolate (lance-shaped; several times longer than wide, broadest at the base, and tapering to a point at the apex). The bloom of the plant is splendid—the floral bracts are wine red, and the flower petals, which flare and recurve, are rose pink. These complement the olive green color of the foliage.

Tillandsia argentina is slow growing, taking years to form a large clump. The plant is one of those species, easy enough to grow and rather inconspicuous in a decent collection, that will manifest its own claim to fame only after a few years have passed. Some knowing wags will buy three or four of this inexpensive species and group them together, saving time and creating the appearance of a clump.

Tillandsia argentina is easier to cultivate outdoors and prefers medium to bright light. This is a hardy plant, because it grows in a wide range of conditions in its natural habitat. When *T. argentina* does grow as a clump, care must be taken to provide sufficient air circulation to allow the inner recesses of the clump adequate opportunity to dry after watering.** If the gas exchange is blocked for a few days, the leaf, leaves, or entire plant, may suffocate and die.

*Supercedes *Tillandsia unca*.

***See* section on Water and Moisture Aspects in Growing Tillandsias, pg. 187.

A.

B.

C.

D. A clump of *Tillandsia argentina* growing saxicolously on a cliff face.

Tillandsia atroviridipetala

Matuda; 1957

(ah-tro-veer-i-di-PEH-ta-la)

subgenus: *Allardtia*

***Atroviridipetala* is a combination of adjectives. *Atro-* is from the adjective *ater*, that means "black." In this case, *atro-* is used as a prefix to indicate a "dark color." *Viridis* is an adjective which means "green," and *petala* means "petals."**

A small, epiphytic or saxicolous Mexican species, *Tillandsia atroviridipetala* occurs at over 3,500 meters in elevation. It is stemless and rarely grows more than 10 cm. in diameter. *Tillandsia atroviridipetala* is graceful with coarsely lepidote, filiform leaves that give the plant a fuzzy, pincushiony appearance.

The inflorescence is compound although difficult to see. The bloom scape is very short, unlike that of *Tillandsia plumosa*, which looks quite similiar but whose scape is much longer and exserts the inflorescence significantly.

The six to eight lepidote, carmine floral bracts are actually borne on individual spikes. The spikes resemble a rooster's comb, which has led to the plant's being termed the "fighting rooster." The small petals are an interesting and unusual fern green.

Tillandsia atroviridipetala is often difficult to cultivate unless the plant is positioned in the same manner as in its native habitat—sideways or upside down. When grown in a vertical position, the meristem often dies when kept moist more than a day or two. Because of its filiform leaves and liberal covering of trichomes, this species thrives under well-lit conditions.

Although *Tillandsia atroviridipetala* is recognized by some as a species, it is not acknowledged by others who put it in synonymy with *T. plumosa*. Renate Ehlers of Stuttgart, West Germany, has collected many intermediate forms of *T. atroviridipetala* and *T. plumosa*. Characteristics such as plant size, leaf shape, trichome size and distribution, phyllotaxis, and habitat diversity provide a fairly constant transition between the two species. Therefore, serious consideration must be given to the idea of synonymy.

However, Sue Gardner (1982)* says that ". . . while apparently close to *Tillandsia plumosa*, *T. atroviridipetala* differs in the length of the scape and in the length and substance of the leaves.

"They also differ in ecological preferences, with *T. atroviridipetala* occurring in arid regions, while *T. plumosa* occurs in humid habitats with frequent fog. The latter species often forms clumps of vegetative offsets, and the former seldom offsets.** It seems preferable to recognize them as different species."

*Gardner, C.S. 1982. *A Systematic Study of Tillandsia Subgenus Tillandsia*. Ph.D. dissertation. College Station. Texas A & M University, pg. 254.

**The author has found that, in cultivation at least, *Tillandsia atroviridipetala* usually produces one or two offsets.

A.

B.

C.

A. A natural branch covered with *Tillandsia atroviridipetala.*

Tillandsia balbisiana

Schultes; 1830

(ball-biss-ee-AH-na)

subgenus: *Tillandsia*

Tillandsia balbisiana **was named for the Italian botanist G. Balbis (1765-1831).**

Tillandsia balbisiana ranges from Florida, the West Indies and Mexico to Colombia and Venezuela. It grows as an epiphyte at elevations from sea level to 1,500 meters and is usually found on forest tree branches.

Tillandsia balbisiana is a stemless, myrmecophytic (ant plant), pseudobulbous species with remarkably long, channeled leaf blades that often recurve, twist, and hang far below the base of the plant itself. The spreading upper blades are often about 40 cm. apart. An average plant is 35-70 cm. high, depending on the length of the decurved blades. The leaves are normally lettuce green, coriaceous (core-ee-AY-shuss; leatherlike, i.e. tough and hard), and covered with finely appressed, cinereous trichomes like *T. caput-medusae* and *T. paucifolia*.

Tillandsia balbisiana produces a tall inflorescence that is blood to cardinal red. The inflorescence usually consists of multiple, complanate spikes that are either densely pinnate (spikes densely packed on opposite sides of a common axis) or subdigitate (almost spreading out from a single point like the fingers of a hand). The inflorescence makes a striking contrast to the comparatively pale color of the leaves. The brightly colored, glabrous floral bracts are imbricate (overlapping) and the corollas are deep royal purple.

I had an interesting experience with a specimen of *Tillandsia balbisiana* a number of years ago. At that time the species name was unknown to me, and the plant was the only one I had ever seen. It was, therefore, unique and valued in my collection. One day I noticed that a colorful bloom spike was emerging from the leaves. After the scape had risen about 25 cm., the plant and its mounting accidently fell and two-thirds of the bloom scape was broken off. Needless to say, I was keenly disappointed. But amazingly, after just a few weeks, I noticed that the scape had started to grow again. Approximately two months later it completed a perfect blooming cycle. The plant then went on to produce numerous pups and there was joy in Mudville!

Because of its coriaceous texture, *Tillandsia balbisiana* is a species that always appears somewhat desiccated—even when watered frequently. The plant prefers bright light, high humidity, and frequent watering when the weather is hot and/or dry. *Tillandsia balbisiana* thrives with about the same care as other myrmecophytic species such as *T. caput-medusae, T. paucifolia,* and *T. pseudobaileyi*.

Tillandsia bergeri

Mez; 1916

(BEAR-ger-ee)

subgenus: *Anoplophytum*

Tillandsia bergeri **was named for a German research scientist, Alwin Berger (1871-1931), who worked with succulents and specialized in *Cactaceae*.**

Tillandsia bergeri is endemic to the province of Buenos Aires, Argentina. It is caulescent and grows saxicolously at elevations up to 100 meters. An average individual plant is about 12-14 cm. wide and 10 cm. high. The leaves are suberect, densely polystichous, and covered with appressed, cinereous scales. The blades are slightly channeled, narrowly triangular, and attenuate. *Tillandsia bergeri* often produces an extensive root system. The roots are thin, branching, and wirelike.

The inflorescence is simple with a number of polystichous, violet and white flowers that superficially resemble those of a miniature iris. The glabrous floral bracts are sometimes a pale carmine.

Tillandsia bergeri is beautiful when flowering, but a common lament is, "Gee, mine never blooms." This is heard often enough to warrant credibility, but the answer to the problem is elusive. Some people feel the plant needs a winter cold spell (my guess, too). Perhaps not enough light is being given to the plant. In any case, there is a missing stimulus that fails to trigger the blooming cycle. However, *T. bergeri* does flower yearly for many growers.

To its everlasting credit, *Tillandsia bergeri* is the "mad pupper" of the *Tillandsia* world. This species grows into a clump more rapidly than any other of the genus when provided with bright light, sufficient water, and fertilizer. *Tillandsia bergeri* is caulescent, and because of its penchant for rapid proliferation, new shoots expand in all directions along the stem. In just a few generations, *T. bergeri* forms a large clump.

As mentioned in the description of *Tillandsia aeranthos*, distinguishing these two species in the non-blooming state is difficult. If myriad pups at all stages of growth are seen growing from among the leaf axils (leaf bases), one may make a safe guess that the plant is *Tillandsia bergeri*. If all a person has to rely on is a single plant, he should believe the name tag. Without a name tag, one must wait for the bloom.

One of the easiest tillandsias to cultivate, *Tillandsia bergeri* thrives under average conditions and tolerates extremes well. It responds with more rapid growth to conditions of bright light, frequent watering, and fertilization. After watering, care should be taken to allow all the leaf surfaces in the clump to dry.

However, one should also watch for dehydration. If the edges of a leaf are touching, or close to it, the plant should be soaked in room temperature water for a day. In this manner the leaf cells will have the opportunity to absorb sufficient water.

A.

B.

C.

Tillandsia brachycaulos

Schlechtendal; 1844
(brah-ki-COW-los)
subgenus: *Tillandsia*

***Brachycaulos* is from the Greek and is a combination of the adjective *brachys,* which means "short," and the noun *caulos,* which means "stem"; hence, it is a reference to the short scape of the inflorescence.**

Tillandsia brachycaulos grows epiphytically in dry woods and deserts in Mexico and Central America at elevations of 600-1,200 meters. It is a widespread species with a number of distinct morphological forms.

One natural habitat is the droughty El Rancho area in Guatemala. *Tillandsia brachycaulos* grows there along with *T. ionantha, T. schiedeana* 'Major', *T. paucifolia,* and *T. xerographica*. In El Rancho, *T. brachycaulos* survives for months at a time without receiving any appreciable precipitation—an indication of a tough, xeric (dry growing) species. However, because its leaves are not as succulent as those of the others mentioned above, *T. brachycaulos* will often dehydrate more rapidly. The plant thrives under high light conditions, and it often blushes a rosy pink when light is combined with a humid environment.

Tillandsia brachycaulos is usually stemless, and the long, attenuate leaves form a spreading rosette that is about 25 cm. across and 12 cm. tall at maturity. The foliage is covered with an indumentum of finely appressed scales.

This species roots readily to a mounting substrate. *Tillandsia brachycaulos* grows best with strong light and commensurate watering. It will grow, albeit more slowly, with lower levels of light. The leaves will become greener as with numerous other species.

The color of *Tillandsia brachycaulos* can vary from moss to silver-green to an almost mauve. But the plant colors dramatically with the onset of the flowering cycle if it is provided with bright light. In just a day or two, the color of the entire plant changes to cherry.

Tillandsia brachycaulos is a species that is pollinated by hummingbirds. From the small floral head nestled in the center of the leaves, deep violet blooms emerge to complement the color dispay. Few species are as rich in color.

One day at the nursery I noticed that a clump of *Tillandsia brachycaulos* was preparing to bloom. Unfortunately, the plants in question were plain green, having been grown in a low light area of the greenhouse. With some trepidation I decided to perform an experiment. I felt that when hung outdoors against the whitewashed greenhouse glass, the glare of the intense summer sun would either cause the clump to become red (if anything could!), or cook it. By the next day the entire clump was fiery red. It did not burn. It was magnificent.

After blooming, *Tillandsia brachycaulos* loses much of its red color. At this point one should soon find two or three pups coming forth, heralding the next generation.

This species is usually more attractive when presented individually rather than in clump form because of the spreading nature of the leaves. In a clump the leaves tend to crowd each other, which makes the appreciation of an individual plant difficult. Additionally, since the plants are competing for space, they will not grow as large.

B.

A. & B. *Tillandsia brachycaulos* in the blooming and nonblooming state.

Tillandsia brachycaulos v. *multiflora*

L.B. Smith; 1945
(mul-ti-FLOR-a)
subgenus: *Tillandsia*

***Multiflora* means "many-flowered."**

Tillandsia brachycaulos v. *multiflora* is native to Guatemala. Although approximately the same size as *T. brachycaulos* v. *brachycaulos*, it is easily distinguished from its varietal twin by its darker green foliage. The dark background also makes the trichomes more visible. These give the plant a velvety look and feel and are responsible for the grayish appearance of the plant. An important, additional difference is that the sheaths are often casteneous (dark brown). When *T. brachycaulos* v. *multiflora* blooms, the plant blushes carmine in contrast to the cherry of *T. brachycaulos* v. *brachycaulos.*

B.

The care for the two forms of *Tillandsia brachycaulos* is basically the same, but *T. brachycaulos* is able to tolerate stronger light than *T. brachycaulos* v. *multiflora*.

A.

A. & B. *Tillandsia brachycaulos* v. *multiflora* in the blooming and nonblooming state.

Tillandsia bulbosa

Hooker; 1826

(bul-BOE-sa)

subgenus: *Tillandsia*

Bulbosus* is an adjective that means "bulbous." The reference is to the swelling of the plant base.

Tillandsia bulbosa has a wide distribution, ranging from Mexico and the West Indies, through Central America, and into Ecuador and northern Brazil. It grows epiphytically on shrubs and trees at elevations from sea level to 1,300 meters. This is a myrmecophytic (ant plant) species with a pseudobulbous base.

I once had occasion to observe this beauty firsthand along the banks of the Rio Dulce on the Atlantic side of Guatemala. The solid green wall of foliage along the river's edge was peppered with brilliant red dots of *Tillandsia bulbosa* in bloom. One could easily see how hummingbirds would be greatly aided in their search for nectar and pollen foodstuffs. The plant is ageotropic (grows irrespective of gravity) and, in nature, is often seen on branches growing horizontally or upside down.

Tillandsia bulbosa is a few-leaved, stemless plant that varies greatly in size. An average adult is about 15 cm. across and 12 cm. in height. In bloom, specimens sometimes grow twice that large. The morphology of *T. bulbosa* is similar to that of *T. butzii* and *T. caput-medusae*. The leaf sheaths are orbicular (more or less round in shape) and the blades are involute and wavy; but unlike the softer leaves of *T. butzii*, they are stiff. The abaxial (outer) surfaces of the blades appear glabrous and green, but they are actually covered with small, widely dispersed and appressed trichomes. The scales of the adaxial (upper or inner) blade surfaces are more pronounced but difficult to see because the blades are greatly involute-subulate (leaf edges greatly inrolled; awl-shaped). The bases have an indumentum of finely appressed trichomes, and the margins of the leaves are amaranth.

Interestingly, the trichomes of *Tillandsia bulbosa* and *T. butzii* vary from other tillandsias in that they change their shape and purpose significantly from one side of the leaf to the other. They are fewer in number on the abaxial leaf surface and are not flexible and hinged like those of the other atmospherics.**

The inflorescence is rarely simple but usually subdigitate. The spikes are spreading, lanceolate, and complanate. The floral bracts are imbricate and covered with finely appressed scales.

When blooming, the inflorescence and upper leaves become cherry. With lavender flowers and fern green, silver, and amaranth sheaths, it is one of the most striking species in the genus.

Tillandsia bulbosa prefers medium light, high humidity, and frequent watering, especially in areas of significant air movement. *Tillandsia bulbosa* is a plant that can be grown as many ferns—frequent watering or misting and medium to low levels of light. As long as adequate moisture is available, high temperatures can be tolerated. It is hardier than most of the other medium to low light species. If positioned horizontally or upside down to help prevent possible rot, *T. bulbosa* would probably grow well as a terrarium plant because of its cultural preferences, size, and shape.

*A characteristic of myrmecophytic plants.

***See* section on Trichome Variations, pg. 208.

A.

B.

C.

B. *Tillandsia bulbosa* growing ageotropically along the banks of the Rio Dulce in Guatemala.

Tillandsia butzii

Mez; 1935

(BUTZ-ee-ee)

subgenus: *Tillandsia*

Tillandsia butzii was described for someone with the surname of Butz, but to date, no information regarding him has come to light.

Tillandsia butzii grows epiphytically in open habitats at elevations of 1,000-2,300 meters. It is found from southern Mexico to Panama. *Tillandsia butzii* has perhaps garnered the title of "strangest looking *Tillandsia*."

The inflated base of the pseudobulb is speckled dark brown and green. The blades are filiform-attenuate, subulate, and from wavy to severely contorted. All of this combines to create an image that almost seems to frighten some souls. Because of its ability to elicit this emotional response, *T. butzii* is a species that many people either enjoy or strongly dislike.

The leaf sheaths are suborbicular (almost round). They often appear nearly glabrous but are actually covered with finely appressed, cinereous trichomes. The sheath margins have large scales that look like so many cilia. An average size adult measures about 35 cm. across between blade tips and 18 cm. in height.

The scape is erect and slender, and the scape bracts are foliaceous (leaflike). The carmine inflorescence is digitate (fingerlike; shaped like an open hand). The bloom spikes are erect to spreading, linear (long, sides parallel), complanate, and thin. The imbricate floral bracts have an indumentum of finely appressed scales. The corollas are violet.

Because *Tillandsia butzii* is not normally covered with a dense indumentum of trichomes, like many similar appearing species, the plant requires more humidity and moisture. Likewise, this species does not need as much light.

Many collectors have experienced the following problem with *Tillandsia butzii*. If they do not water their plant often, it rapidly desiccates; if they do water the plant frequently, it seems to rot—an apparently "damned if you do, damned if you don't situation."

By cultivating the plant like *Tillandsia atroviridipetala*, in a horizontal or upside-down position, one can water it frequently without having to worry about water collecting inside the pseudobulb at the meristem of the plant—a probable cause of the rot. *Tillandsia magnusiana* Wittmach, *T. matudae* L.B. Smith, *T. oaxacana* L.B. Smith, *T. macdougallii* L.B. Smith, and *T. schiedeana* are others that should be mounted similarly.

This is a difficult plant to maintain indoors, let alone grow—usually because of a lack of humidity. Outdoors in a patio situation or under a tree where *Tillandsia butzii* receives sufficient humidity and moisture, the plant will flourish. Experience has also shown that *Tillandsia butzii* is much easier to grow once it has developed into a clump. At this stage it is almost indestructible. *Tillandsia butzii* is a good companion plant for *T. bulbosa, T. punctulata,* and *T. filifolia*. This species grows well in conditions of medium to low light combined with frequent watering.

A.

B.

C.

Tillandsia cacticola

L.B. Smith; 1954

(kahk-TI-coe-la)

subgenus: *Phytarrhiza*

***Cacticola* is a combination of the noun *cactus* and the adjectival ending *-cola*. Now the name of a new world family of plants, *cactus* in classical Latin meant a "prickly plant with edible stalks," or "anything thorny" or "unpleasant." *-Cola* means "dweller," so the epithet stands for "cactus dweller."**

Tillandsia cacticola grows in dry locations in northern Peru on shrubs, cacti, and trees such as acacias and erythrinas. It occurs at elevations of 300-2,300 meters. Spectacular when in bloom, this species is also easy to grow.

A mature plant is normally about 25 cm. wide and 10 cm. high. *Tillandsia cacticola* is stemless, with narrowly triangular, involute blades that form a spreading rosette. The lack of a stem and the characteristic shape of the plant easily differentiate it from two species with which it is sometimes confused, *T. straminea* and *T. purpurea*. When well grown, the downy indumentum of trichomes and the symmetrical shape of the plant make it a most attractive species.

B.

However, it is the bloom of *Tillandsia cacticola* that has justly won its high standing in the eyes of *Tillandsia* fanciers. The tall, slender, 25 cm. scape terminates in a number of spikes that are usually intense lavender when mature. The size and brilliance of the lavender spikes can be seen from a great distance. The spikes are large, broadly elliptic (broadest at the center, tapering equally to the base and apex) and complanate.

The glabrous floral bracts are usually imbricate with no visible space between successive nodes. This is what gives the spike its unified appearance. The imbricate floral bracts are in contrast to those of *T. straminea* and *T. purpurea*, which have internode space among their tubular, not complanate, bracts.

The petals vary from cream to pale yellow and have recurved tips that are violet. Although some have said the flowers are fragrant, I have yet to encounter one that is scented.

The plant should receive bright light with enough moisture to keep it looking lush. Fertilization throughout the warm months will produce beneficial results. Although a strong, robust plant, *Tillandsia cacticola* does not often produce more than one or two pups. Therefore, it is not a species that grows quickly into a large clump. Once *T. cacticola* is permanently situated in an ideal spot, it will attach to its substrate with large roots, anchoring the plant for generations to come.

A.

B. The inflorescence of *Tillandsia cacticola* 'Splendid Bloom.

Tillandsia caput-medusae

Morren; 1880

(KAH-put meh-DOO-sigh)

subgenus: *Tillandsia*

***Caput* means "head" and *medusae* means "of Medusa." In Greek mythology, Medusa was the Roman Gorgon whose *coiffure* consisted of wriggling snakes—an imaginative metaphor for the leaf blades of this species.**

Tillandsia caput-medusae is one of the most widespread and hardy species. It grows epiphytically from sea level to 2,400 meters in elevation and is indigenous to Mexico and Central America.

Tillandsia caput-medusae is stemless and quite variable in size. In Guatemala, small forms rarely more than 10 cm. high thrive in dry gullies that are virtually cloudless and rainless except for the July-August rainy season. Humidity from trickling streams and gullies allows the plants to exploit the habitat. The intense heat and lack of available water keep the plants from attaining a larger size.

Larger forms that are about 24 cm. wide and 20 cm. high also grow in droughty habitats but receive moisture at more frequent intervals. Some giant specimens of this species have been known to reach a height and width of 35 cm.

Most populations are lettuce green, but a few forms are much darker. These have very long leaves with blades that are often three times the length of the sheaths. This form of *Tillandsia caput-medusae* is rare in cultivation.

The sheaths often have a colorful beet purple margin. They are broadly ovate (egg-shaped and attached at the wide end) and inflated, forming an ovoid pseudobulb. This myrmecophytic characteristic allows a great amount of space between each successive leaf layer. The sheaths merge gradually into the long, attenuate, involute blades that are often contorted. The leaves have an indumentum of coarse, cinereous trichomes that are especially prominent on the abaxial (outer) surface of the leaf bases. The blade tips are subulate.

In rare cases the inflorescence is simple, but normally it is digitate with two to six spikes. The spikes are erect to spreading and from straight to curved. The shape is linear-lanceolate (long, thin, and flat and pointed at the end).

The floral bracts are imbricate, chartaceous (papery), and usually glabrous, much like those of *Tillandsia bulbosa*. Most other myrmecophytic species have lepidote (prominently scaled) floral bracts. With strong light, the floral bracts are often a strikingly attractive cherry to blood red. Sometimes they are a combination of red and green and, once in a while, lettuce green. The amount of light the inflorescence receives plays an important part in the color of the bloom spikes. The corollas are mauve.

Tillandsia caput-medusae tolerates poor conditions and neglect better than most other species. In addition, the plant responds well to good care. A program of high light, water, and fertilizer will reward the grower with more rapid plant growth, a larger inflorescence, and more offsets.

When grown well the plant is quite colorful. The leaves are covered with a velvety, silver indumentum while the margins are often bright purple. *Tillandsia caput-medusae* roots rapidly to its mounting medium. In just a few years one will be able to point proudly to a specimen clump 60 cm. in diameter that truly appears as Medusa's head!

A.

B.

Tillandsia concolor

L.B. Smith; 1960
(CONE-co-lor)
subgenus: *Tillandsia*

Concolor* is an adjective that means "of one color," a reference to the uniform leaf color.

Tillandsia concolor is a hardy, stiff-leaved species with an altitude range of 150-1,200 meters. It grows epiphytically in exposed habitats in Mexico and El Salvador. *Tillandsia concolor* is highly adaptable to and tolerant of different care situations.

Tillandsia concolor forms a stemless, spreading rosette that can grow to 30 cm. in diameter and 18 cm. in height. Besides its sometimes similar appearance to *Tillandsia fasciculata, T. concolor* resembles *T. ixioides* in the non-blooming state. They are easy to distinguish because the narrowly triangular, involute leaves of *T. concolor* are dull, olive green, while those of *T. ixioides* are more lepidote and a gray, silver-green. The leaves are covered with an indumentum of subappressed, cinereous scales.

The plant often incurs minor leaf damage when in transit because of its brittle leaves. For those who understand this, a couple of broken leaves is not an important factor in choosing a plant because *Tillandsia concolor* grows rapidly with good conditions.

The inflorescence is either simple or digitately compound. The spikes are lanceolate and slightly inflated. Among individual plants, the imbricate, slightly inflated floral bracts vary from green to cherry. Most often the bracts are a combination of the two, with the margins coloring red and the rest of the bract remaining green.

The petals of *Tillandsia concolor* are ruby red. No other xeric *Tillandsia* has the combination of this exquisite color and large corollas. With numerous large spikes, one is assured of many blooms over a period of weeks. *Tillandsia concolor* also produces numerous offsets.

Tillandsia concolor flourishes under a myriad of conditions—from very high light levels to shade. Watering should be consistent with the amount of light the plant receives. Bright light combined with a fertilization program in the spring, summer, and fall will speed growth, produce a larger inflorescence, and encourage more offsets.

*In nature, populations of *Tillandsia concolor* and *T. fasciculata* are sympatric, and a key way to distinguish them when not in bloom is by the sheath color. *Tillandsia fasciculata* sheaths are dark ferruginous (rust-colored) or casteneous.

A.

B.

Tillandsia crocata

Baker; 1887

(crow-KAH-ta)

subgenus: *Phytarrhiza*

***Crocatus* means "with saffron yellow." The first part of this adjective, *croc-*, is from the word *croceus*, which means "saffron yellow." The suffix, *-ata*, indicates "possession" or "likeness."**

Tillandsia crocata is a small, clumping species that grows saxicolously in Brazil, Argentina, and Uruguay at elevations of 900-2,700 meters. The brilliant color and surprisingly strong, sweet fragrance of the flowers are unsurpassed in a small species and warrant its great desirability among collectors.

The plant may be simple or few-branched, and the leaves are distichous (in opposing rows). Mature plants average 10-15 cm. in height and width. The leaves are attenuate, involute, spreading, and slightly recurved or reflexed. The blade tips are subulate.

The leaves are densely covered with coarse scales, the wings of which are large and elongated on one side. By rising up from the leaf surface, the wings aid in the epidermal drying process by facilitating greater air movement.* This faculty is essential for clumping species that receive water frequently at certain times of the year.

The slender bloom scape subtends the simple, distichous inflorescence. The densely lepidote floral bracts are imbricate. There are at least two types of *Tillandsia crocata*—one that produces canary yellow flowers, the other buttercup yellow.

Tillandsia crocata is easy to cultivate, and it is drought tolerant. This species prefers bright light because of its dense trichome tomentum. In a dark location the plant will gradually weaken and wither. In addition, because of its small size, *T. crocata* cannot survive lack of water or humidity for as long as more succulent and larger-leaved species such as *T. duratii.*

All of the leaf surfaces should dry between waterings. If the plants in a clump become wet and remain so for a few days they may suffocate and die. In a clump, air cannot circulate among the leaves as well as in individual plants. However, this is a tough species, and it will thrive within a broad latitude of care conditions. *Tillandsia crocata* will become a featured favorite every blooming season once a specimen has grown for a few generations and developed into an attractive clump.

**See* subsection on Trichome Variations, pg. 208.

A.

B.

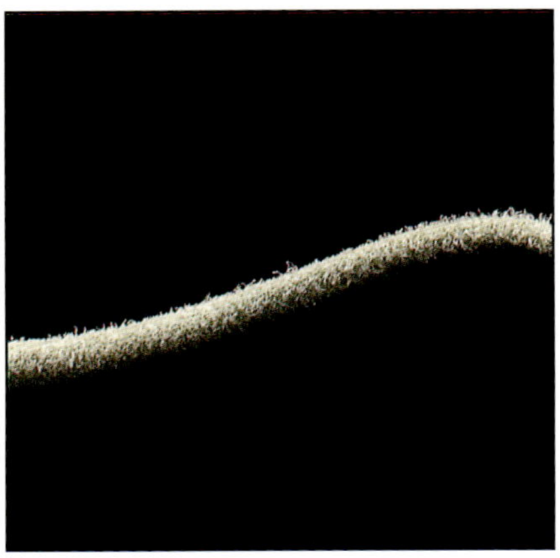

C.

B. Note lighter colored pollen grains on petal blades.

C. Photo detail of coarse trichome tomentum on leaf blade.

Tillandsia cyanea

Linden; 1867

(sigh-AH-nay-ah)

subgenus: *Phytarrhiza*

***Cyaneus* is an adjective originally from the Greek that means "dark blue," a reference to the color of the flower petals.**

Tillandsia cyanea grows epiphytically in the forests of western Ecuador at elevations from near sea level to 850 meters. In bloom, it is one of the most spectacular species, and it has been produced commercially for generations in Europe because of this bloom. This is one of the few species that grows more robustly as a terrestrial.

Tillandsia cyanea is stemless and about 35 cm. in width and 18 cm. in height in the non-blooming state. The many fern green leaves form a spreading rosette. They are linear-triangular, attenuate and recurve gracefully. The bases are often cardinal, and the green blades have longitudinal, cardinal striations. The foliar trichomes are so small and widely dispersed that the leaves appear glabrous to the naked eye.

The simple, glabrous inflorescence is distichous and usually about 10-15 cm. in height and 5-6 cm. in width—which is relatively large and showy for a *Tillandsia*. It is elliptic and complanate, rather like a paddle. In medium to bright light this large inflorescence becomes bright pink. The floral bracts are densely imbricate and distichous. The beautiful violet petals are spreading and large—usually 2 cm. across and 3.5 cm. in length. As with most species, each floral bract gives rise to a bloom that will last for a day or two. This means the plant will be in bloom for as long as a month. Sometimes the flowers have a cinnamon fragrance.

Tillandsia cyanea normally produces from three to a dozen offsets, not only during and subsequent to the blooming period but during the growing cycle as well.

Tillandsia cyanea is easy to cultivate, but it is normally grown in more shaded and protected conditions than other xerophytic species. This plant should be potted in a fast-draining mix and watered frequently during the growing season. It makes a fine houseplant if given sufficient light and proper watering.

Tillandsia cyanea responds favorably to a consistent feeding program. The soil should be fertilized in the same manner as that of any regular houseplant. For those who desire, *T. cyanea* can be mounted epiphytically in lieu of life in a pot. A swath of sphagnum wrapped around the root base helps keep the roots from drying. The moss should not be kept dripping wet, but neither should it dry completely.

At least during the warm months, "Grow the plant outdoors, and enjoy the bloom indoors," seems to be a good motto for this species—and for most others as well.

A.

B.

C.

Tillandsia diaguitensis

Castellanos; 1929

(dee-a-gwee-TEN-sis)

subgenus: *Anoplophytum*

The Indian name for the type locale of this species is the pueblo Diaguitas, Argentina. *-Ensis* is an adjectival suffix that indicates origin or place.

Tillandsia diaguitensis is native to Argentina and Paraguay and grows at elevations from 200-1,600 meters. The countryside surrounding one habitat area in the foothills of the Argentine Andes is some of the most breathtaking I have ever seen. The tapestry of landscape colors reminded me of southern Utah and the national parks of Bryce and Zion. There are few people, fewer roads, and the weather is usually temperate.

Tillandsia diaguitensis grows as a saxicole on mountainous rock outcroppings and their corresponding arroyos. It is also found epiphytically in the same areas, growing on small shrubs and in cactus thickets. *Tillandsia diaguitensis* averages 9 cm. wide and

B.

12-25 cm. tall. It is caulescent with a long, rigid stem. Once mounted in a given direction, *T. diaguitensis* will grow toward the light source.

The leaves of *Tillandsia diaguitensis* are silver-green, lepidote, and prominently veined. These traits are shared, more or less, with the sympatric *T. xiphioides, T. peiranoi,* and *T. muhrii*. The leaves are also polystichous, lax (not dense) and stiff. The blades are erect to recurved, narrowly triangular, attenuate, and involute-subulate.

The inflorescence is simple and lanceolate. It is similar to those of *Tillandsia xiphioides* and *T. xiphioides* v. *tafiensis*, although it is longer and more slender. The floral bracts are moss or olive green to beige, imbricate, and glabrous. They produce large, sweetly pungent white flowers. The floral bracts may wither or the flowers may not develop completely if the air is dry. For this reason, should the plant begin to produce an inflorescence, it should be provided substantial humidity.

The plant does not grow as quickly or clump as fast as *Tillandsia bergeri,* but it does produce offsets continuously from among the leaf axils during the growing season. *Tillandsia diaguitensis* grows well with high light and humidity levels, although it is hardy and will flourish with less than ideal conditions.

A.

B. *Tillandsia diaguitensis* growing in clumps of *Deuterchonia,* another genus of bromeliad.

Tillandsia distícha

Humboldt, Bonpland, & Kunth; 1816
(DISS-ti-ka)
subgenus: *Allardtia*

***Distichus* is an adjective borrowed from the Greek. It means "arranged in two opposite rows," a reference to the floral bracts.**

Tillandsia disticha grows saxicolously or epiphytically in dry woodland habitats in Peru, Ecuador, and Colombia. It occurs from near sea level to 2,200 meters in elevation. This species shows substantial variation among half a dozen subspecies. A large population from Ecuador normally has a purplish cast to the leaves, although the plant often loses some of its color in cultivation. It is called *T. disticha* 'Major', and it is the most common form in cultivation.*

Mature specimens of *Tillandsia disticha* 'Major' are about 30 cm. across and 50 cm. tall. The plant differs from *T. disticha* v. *disticha* in two significant ways—the scape of *T. disticha* is shorter than the leaves, and the offsets do not grow at the ends of stolons (runners). The scape of *T. disticha* 'Major' exceeds the leaves, and the plant is stoloniferous (produces an offset at the end of a runner).

Tillandsia disticha is stemless with polystichous leaves that form a globose (spherical) pseudobulb. The sheaths are broadly ovate or orbicular and contract abruptly, more or less, into the blades. They

D.

often possess a number of longitudinal wrinkles. The blades are involute-subulate and filiform-attenuate. The epidermis of *T. disticha* 'Major' is prominently nerved with an indumentum of appressed, cinereous trichomes.

The bloom scape is erect and slender. The digitate, complanate spikes are suberect, linear, and lettuce to moss green. The glabrous to subglabrous floral bracts are distichous and imbricate. The canary petals are small with narrowly lanceolate blades. The flowers sometimes have a slight, sweet fragrance.

There is another, smaller form, also from Ecuador, that is much more glabrous and green. It has large, cinereous, ciliate scales along dark maroon leaf margins. The color contrast among the green, purple, and white is striking. This form is known as *Tillandsia disticha* 'Minor.'* Adults are about 16 cm. across at the leaf tips and 26 cm. tall. Also, the more lepidote floral bracts of this form have margins that are often tinged with maroon.

Tillandsia disticha 'Minor' grows in trees where light levels are lower and precipitation is more frequent. This is reflected in the greener appearance of the plant. *Tillandsia disticha* 'Minor' is hardy, tolerating bright light and dry periods.

The filiform blades are a sign that *Tillandsia disticha* does not grow well indoors due to the usual lack of light, humidity, and air movement. Rather, it thrives outdoors in bright light. It requires high humidity and/or frequent watering in periods of high temperature or aridity. In a good location that meets these conditions, *T. disticha* is a tough, hardy plant.

Tillandsia disticha is appreciated by many, for as a clump develops over several seasons the individual plants space out from one another so each plant can be better seen in its entirety. Also, each plant can grow to its full potential without being encumbered by others in the group.

*First use of this name.

A.

B.

C.

B. *Tillandsia disticha* 'Major' showing grayer, more lepidote leaves.

C. Base of *Tillandsia disticha* 'Minor' showing greener leaves with ciliate, purple margins and characteristic slight wrinkling of sheaths.

D. Lichen on cactus indicates constant humidity.

Tillandsia duratii

Visiani; 1840

(dur-AH-tee-ee)

subgenus: *Phytarrhiza*

***Tillandsia duratii** was named for an Italian, Durat, who was the first to cultivate this species in Europe.**

Tillandsia duratii grows as an epiphyte primarily in the drier areas of western Bolivia, eastern Paraguay, and northern Argentina, known as the Great Chaco. It is also abundant in the Andean foothills of Argentina and southern Brazil. In terms of sheer numbers and habitat range, *T. duratii* is one of the most successful *Tillandsia* species. It occurs at elevations of 200-3,500 meters.

There are two forms of *Tillandsia duratii—T. duratii* v. *duratii* and *T. duratii* v. *saxatilis* L.B. Smith (sahks-AH-til-is). The chief distinguishing characteristic between the two forms of *T. duratii* is the shape of the inflorescence. The spikes of *T. duratii* are erect and strict (tight against the rachis), while the spikes of *T. duratii* v. *saxatilis* are curved-spreading.

Native populations of *Tillandsia duratii* often vary greatly in size. Small mature plants often bloom when about 20 cm. in width and height. Giant specimens grow to 40 cm. across and a meter high, not including the impressive inflorescence that is often another 80 cm. The stem on a specimen such as this may be 4 cm. in diameter.

Tillandsia duratii is caulescent. The polystichous leaves are long, narrowly triangular, and involute. They are densely covered with subappressed, cinereous trichomes.

The most interesting aspect of this species, besides its magnificent bloom spike, is the remarkable way in which the leaves tightly recurve. In nature, *Tillandsia duratii* almost always depends on these recurving leaves for growing support. As the older, lower leaves die back they desiccate, shrinking into tight ringlets around a convenient twig or branch. The plant grows upward along with the tree, pushing its way through the leaves and branches and maintaining its position of exposure to the sun. *Tillandsia duratii* sometimes covers much of a tree, appearing as a flock of resting white egrets.

The inflorescence of *Tillandsia duratii* develops for several months, after which it produces heavily fragrant blooms for another month or so, longer than any xeric species. The linear to lanceolate spikes are complanate. The imbricate, distichous, sublepidote to subglabrous floral bracts are stamineous and fern green. Each flower has three large, mostly lilac petals. The throat of the corolla is white. For plant size, shape, and durability in combination with the bloom size and fragrance and longevity of its blooming cycle, *T. duratii* is in a class by itself.

Few *Tillandsia* species are easier to cultivate than the two forms of *T. duratii*. They thrive under a wide range of light, water, and temperature conditions. The water and light levels should be complementary, and when given in high doses, *T. duratii* often grows very large.

*Supercedes *Tillandsia decomposita*.

A.

B.

C.

A. *Tillandsia duratii* v. *saxatilis* growing epiphytically in Bolivia.

B. Detail showing trichomes on floral bracts.

D.

D. *Tillandsia duratii* v. *saxatilis* in bloom. For size comparison, note *T. usneoides* and blooming *T. albertiana* at intersection of two branches.

E.

F.

G.

E. *Tillandsia duratii* v. *duratii;* note spikes tight against rachis.

G. *Tillandsia duratii* v. *saxatilis* growing on scrub bushes in the Andean foothills of Argentina.

Tillandsia edithae

Rauh; 1974

(E-dith-eye)

subgenus: *Allardtia*

This species was named for the wife of Alfred Blass of West Germany who first bloomed the plant.

Tillandsia edithae grows on rocks and cliff faces at altitudes of 2,700 meters near Sorata, Bolivia. *Tillandsia edithae* is popular among collectors because of its brilliantly colored inflorescence.

The area to which *Tillandsia edithae* is native has some of the most breathtaking scenery in South America. High Andean peaks surround verdant, chiseled valleys. The infrequent towns along the precipitous mountain roads cling precariously to the sides of these majestic *cordilleras*. One has to be just a little envious of the Bolivians, including the many colorfully dressed Indians, who call this free and pristine land home.*

Tillandsia edithae is caulescent with many densely polystichous leaves. The blades are more triangular than those of many other species, and they are covered with gray, subspreading scales. A mature specimen of *T. edithae* is approximately 12 cm. in diameter and 10-15 cm. in length. The plant roots quickly with good growing conditions.

C.

The simple inflorescence is thick and capitate (forming a rounded head). The 8-12 subglabrous floral bracts are crimson, and the polystichous corollas are cherry.

Tillandsia edithae produces both acrotonal (at the base of the inflorescence) offsets and adventitious or hair pups along the stem. These grow more slowly and often separate from the parent plant because they are weakly attached.

When mounted on wood in a hanging position, *Tillandsia edithae* becomes more caulescent with each succeeding generation. The plants eventually form a dense clump. This species lends itself to placement on an appropriately colored seashell because of the shape of the plant and its crimson bloom.

Tillandsia edithae grows well with bright light, water, and a scheduled feeding program. The plant will thrive as long as it is not grown in damp, dark conditions.

*See photo, pg. 106.

A.

B.

A. Photo of bloom enlarged approximately four times to show detail.

B. Unforeseen circumstances often hinder or foil even the best laid plans.

Tillandsia fasciculata

Swartz; 1788

(fah-sih-coo-LAH-ta)

subgenus: *Tillandsia*

***Fasciculatus* is an adjective derived from the noun *fasciculus*, "a small bundle." *-Ata* indicates possession. *Fasciculata* means "clustered" or "growing or gathered in bundles," a reference to the phyllotaxis (arrangement) of the leaves.**

Tillandsia fasciculata varies greatly in habit—Smith and Downs (1977) list 10 separate varieties in their monograph.* Recently, however, some of these classifications have been changed because a few of the varieties are valid species while others have been reduced to synonomy.** It is beyond the scope of this work to present separately the vegetatively similar *T. rotundata* L.B. Smith, *T. jaliscomonticola* Matuda, *T. rodrigueziana* Mez, and the other varieties of *T. fasciculata*. For the purposes of the following discussion, all of these species and varieties may be grouped together.

Tillandsia fasciculata and closely related species occur from Florida, through the Caribbean and Central America, and into northern South America. They are found growing epiphytically from sea level to 1,800 meters.

The plants form stemless rosettes that are 20-100 cm. high. The sheaths are dark ferruginous (rust-colored) or castaneous. The blades are narrowly triangular, attenuate, and 30-70 cm. long. The leaves are stiff, brittle, and easily broken if bent. They can be erect, spreading, or even decurving.

The scape of the inflorescence is tall and erect in most varieties. The inflorescence can be simple but is often digitately compound. The spikes are usually erect to spreading and complanate. In many varieties, the floral bracts are imbricate, glabrous to subglabrous, and distichous. They are often a shade of fern green with cardinal to blood red margins. The corollas are mauve to indigo.

The different forms of *Tillandsia fasciculata* are easy to grow. They prefer bright light and fresh air, and although they may be classed as a xerophytes, frequent watering and fertilizing will promote faster growth. If the base of a plant is allowed to hold water for an extended length of time, it could become stagnate. Stagnant water serves as host for rot-causing forms of fungi and bacteria.

Tillandsia fasciculata is prolific, often producing a half dozen offsets. As these new plants develop they will begin to crowd each other. They should be separated at this point and mounted individually to maximize the growth rate and plant conformation. With sphagnum moss wrapped around the root base, the plant will soon develop an intricate, holdfast root system. If grown as a potted plant in a well-draining medium, the plant will soon fill the pot with roots.

Because *Tillandsia fasciculata* often grows into a large specimen, it can serve as the centerpiece in one's collection. When allowed to grow for a period of time, a plant will produce many attractive, symmetrically arranged leaves along with a sizeable and long lasting colorful inflorescence.

*Smith, L.B. and Downs, R.J. 1977. *Tillandsioideae in Flora Neotropica. Monograph No. 14. Part 2.* New York, N. Y. Hafner Press.

**Gardner, C.S. 1982. *A Systematic Study of Tillandsia Subgenus Tillandsia.* Ph.D. dissertation. College Station. Texas A & M University.

A.

B.

C.

A. *Tillandsia fasciculata* v. *fasciculata.*
B. This specimen of *Tillandsia fasciculata* v. *densispica* was grown in a pot. Note the extensive root system.
C. Inflorescence of *Tillandsia fasciculata* v. *rotundata.*

Tillandsia filifolia

Schlechtendal & Chamisso; 1831

(fi-li-FO-lee-a)

subgenus: *Tillandsia*

***Filifolius* is a combination of the noun *filum*, which means "thread," and *folius*, the adjectival form of *folium*, the word for "leaf." *-Folius*, means "leaved" when used in combination with another word.**

The range of *Tillandsia filifolia* extends from Central Mexico to Costa Rica at elevations of 100-1,300 meters. It is a mesic species that grows epiphytically in areas that receive medium to low light and substantial moisture. This species occurs singly or in large mats on tree branches. The plant is also ageotropic.

Tillandsia filifolia can be easily confused with *T. argentea*. Both species look like pincushions with full complements of pins, although the blooms of the two are quite different.* *Tillandsia filifolia* is often stemless, and the leaves form a dense rosette. The blades of *T. filifolia* are fern green with grape violet to castaneous sheaths that are frosted with small, widely dispersed, subappressed trichomes.

The inflorescence is bipinnate with many slender, suberect to spreading spikes. The small, nile green and maroon floral bracts are glabrous and distichous. The petals are spreading and aster violet.

Tillandsia filifolia normally produces numerous offsets that separate easily by hand when about one-half to two-thirds adult size.

Tillandsia filifolia is usually acquired when about 20 cm. in diameter and 10 cm. tall. It can be grown much larger with good cultivation. The plant should be watered frequently in dry climates and moderately in those that are humid because it prefers cool conditions and substantial fresh air movement. Because of the filiform leaf blades, which is an indication of air movement preference, it does not normally flourish indoors. However, *T. filifolia* thrives in a patio situation, especially with other plants close by to add humidity.

Tillandsia filifolia is a popular species among collectors because of its unusual color combination and graceful shape.

**See* description of *Tillandsia argentea* for more discussion.

A.

B.

C.

Tillandsia friesii

Mez; 1906

(FREE-see-ee)

subgenus: *Anoplophytum*

Tillandsia friesii **was named after Fries, who was the first person to collect the plant.**

Tillandsia friesii grows saxicolously in clumps from 3,000-3,550 meters in the Andean valleys of northwestern Argentina. Mature, non-blooming plants average about 12 cm. in width and 9 cm. height. *Tillandsia friesii* is a caulescent species with densely polystichous, cinereous and coarsely lepidote leaves. The blades are faintly nerved, narrowly triangular, and attenuate. Scales on the blade margins are ciliate (fringed with conspicuous hairs along the margin).

The simple, glabrous inflorescence equals or nearly equals the leaves in length and is distichous, sublanceolate, and complanate in shape. The dull, wine red floral bracts are imbricate. The corollas are rose and usually number less than six. The flared petal blades are recurving.

Tillandsia friesii is similar in appearance to *T. muhrii* but is distinguished from it by more prominently scaled leaves. The morphology also differs. The blades of *T. friesii* spread uniformly and recurve slightly; the older leaves of *T. muhrii* spread and recurve greatly while the newer leaves are suberect. The scape of *T. muhrii* exserts the simple bloom spike beyond the leaves, and the blooms are indigo.

Tillandsia friesii grows in areas where some clumps are exposed to direct sun for a few hours of the day. The area high in the Andes to which it is endemic has a relatively low level of precipitation. The only other species that grows in close proximity is *T. capillaris* v. *virescens* (Ruiz & Pavon) L.B. Smith—probably the smallest *Tillandsia*. Other, miniature species in subgenus *Diaphoranthema*, along with *T. xiphioides*, grow many kilometers away at lower elevations in the *quebradas* (ravines).

Tillandsia friesii is a durable species that tolerates a wide range of care conditions. It prefers bright light and will grow faster with frequent watering and fertilizing in dry climates. Its gracefully arched, silver white leaves make it a most attractive plant.

A.

B.

Tillandsia incarnata

Humboldt, Bonpland, & Kunth; 1816

(in-car-NAH-ta)

subgenus: *Allardtia*

Incarnatus **is an adjective that means "flesh-colored," often with a slight reddish cast. This refers to the variable but usually pale pink to rose inflorescence.**

Tillandsia incarnata grows terrestrially, epiphytically, and saxicolously in Colombia and Ecuador at elevations of 500-3,200 meters. It is one of the few species that thrives in all three growing environments.

There is a small, ephiphytic form of *Tillandsia incarnata* that clumps, forming a round ball. It is quite attractive when suspended from a piece of wire. The other, more common form, is more caulescent and grows saxicolously near Quito, Ecuador. The plants are found hanging in large clumps on cliff faces. The growing tip of the plant recurves and grows upward toward the light source. The stems branch as they grow, forming ever larger clumps until the weight of the whole is too much to support it. At this point, at least some of the clump breaks away, falling to the base of the cliff.

A mature plant is 10 cm. across, and although the length varies greatly, a nice specimen is about 40 cm. long. The leaves are soft and narrowly triangular. They are polystichous and covered with cinereous, appressed, or slightly spreading scales. The blade tips are often involute.

The bloom scape is long and slender, growing erect or curving slightly. The scape is red but becomes a pale rose because of the indumentum of trichomes. The bloom spike is simple; the densely lepidote floral bracts are imbricate and distichous; the small petals are rose.

Tillandsia incarnata grows rapidly with bright light, water, and feeding. It is an outdoor species, preferring strong air movement. It also dehydrates quickly from transpiration because the leaves are less succulent than those of many other xeric species.

Therefore, it must be watered frequently in dry environments. Outdoors in a well lit location, *Tillandsia incarnata* will form a large clump in just a few growing seasons. After the clump has developed, the main problem confronting the grower will be keeping the clump together because of its loose attachment where the individual plant divisions occur.

Tillandsia ionantha

Planchon; 1855

(ee-oh-NAHN-tha)

subgenus: *Tillandsia*

Ionanthus **is from the Greek and is a combination of the adjective, *ion*, which means "violet," and the noun, *anthos*, which means "flower."**

Tillandsia ionantha v. *ionantha* grows primarily as an epiphyte, but, according to Smith & Downs, it is also found as a terrestrial.* It grows at elevations of 450-1,700 meters from Mexico to Nicaragua. The ubiquitous *T. ionantha* is a most popular species and is often the first *Tillandsia* one acquires. It is attractive, inexpensive, small, and easy to cultivate. In the natural environment, *T. ionantha* is often exposed to the full intensity of the elements so it is particularly adapted to severe xeric conditions.

Tillandsia ionantha is a stemless plant with 20-40 succulent, polystichous leaves that form a dense rosette. *Tillandsia ionantha* often grows in groupings of three to five plants. Sometimes it develops into large clumps that may completely encircle a tree branch.

The leaves are narrowly triangular, subulate-attenuate, and erect. The abaxial sheaths are

D.

covered with dense, coarse, cinereous trichomes. The leaves are lettuce to fern green and sometimes tinged with crimson. This may be part of the plant's genetic makeup, a result of bright light and humidity, the approach of blooming time, or death.

To make sure the plant is healthy, the centermost leaves should be tugged gently. If they remain intact, the plant is fine. Should they pull out easily, and if the bases of the pulled leaves are brown or black, continue removing the centermost leaves until resistance is felt. What remains should be allowed to dry for a few days, after which the plant can be cultivated in the usual fashion. This will probably result in a number of offsets.

Tillandsia ionantha produces abundant roots when grown well. They attach the plant firmly to the mounting substrate. When cultivated in bright light, the entire plant changes in a short time to a fiery crimson that commands the spotlight. The grape violet flowers emerge from the center of the plant on abbreviated spikes deep within the plant.

The variety of *Tillandsia ionantha* from Guatemala varies significantly from its Mexican neighbor. Individual adult Guatemalan plants have fuzzier leaves and are 8-10 cm. in height and width. Specimens of the Mexican variety are smaller and generally 5-6 cm. in height and width.

Tillandsia ionantha is a durable species. With its attractiveness and size, the plant is ideal to locate on a kitchen windowsill, where it can receive strong light and be observed frequently.

*Smith, L.B. and Downs, R.J. 1977. *Tillandsioideae in Flora Neotropica. Monograph No. 14 . Part 2.* New York, N.Y. Hafner Press, pg. 913.

A.

B.

C.

A. A well-rooted specimen of the Guatemalan form of *Tillandsia ionantha*.

B. Mexican form of *Tillandsia ionantha* in bloom.

C. An average, nonblooming Guatemalan specimen; note the secund habit of the leaves.

D. *Tillandsia ionantha* growing on branches in the El Rancho region of Guatemala.

Tillandsia ionantha v. *scaposa*

L.B. Smith; 1941

(ska-POE-sa)

subgenus: *Tillandsia*

***Scapus* is from the Greek and means a "shaft," "stem," "stalk," "trunk," etc. In botanical usage, *scaposus* means "having a well developed scape."**

Tillandsia ionantha v. *scaposa* grows epiphytically in the forests of Guatemala at altitudes of 1,500-1,950 meters. It is distinguished from *T. ionantha* by its short but distinct bloom scape and distichous, imbricate floral bracts.

Additionally, vegetative differences are quicky ascertained by the experienced eye. Of the number of forms of *Tillandsia ionantha, T. ionantha* v. *scaposa* is the largest. Plants of 15 cm. in width and height are not uncommon, although average mature plants are closer to 10 cm. The phyllotaxis is not as dense as the other forms. The leaves of *T. ionantha* v. *scaposa* are less stiff and softer than those of *T. ionantha* v. *ionantha*. Also, the foliage is slightly different in color, that is a darker and grayer green.

When in bloom, the upper part of the plant becomes carmine. The rest of the plant remains the same gray-green. With *Tillandsia ionantha,* the entire plant changes to crimson when given strong light. The corolla of *T. ionantha* v. *scaposa* is mauve to amethyst rather than grape violet.

Tillandsia ionantha v. *scaposa*

D.

E.

does not tolerate as much light or survive dry periods as long as *T. ionantha*. The leaves lose internal water via transpiration more quickly, so care must be taken in drier climes whether indoors or out. *Tillandsia ionantha* v. *scaposa* should be watered more frequently and/or given more humidity than *T. ionantha* v. *ionantha*. This plant is more robust when grown at cooler temperatures.

Other Forms of *Tillandsia ionantha*

There are other forms of *Tillandsia ionantha* besides the two previously mentioned. *Tillandsia ionantha* v. *vanhyningii* Foster is caulescent and grows on vertical cliffs of limestone in Chiapis, Mexico. Generally, the leaves are thicker and more succulent than the other varieties.

Tillandsia ionantha 'Rosita'* is a beautiful, uncommon Mexican form that is red throughout its life. It is the size and shape of the standard Mexican form, although the blade tips are sometimes more erect. Interestingly, second generation plants often grow much larger—up to 10 cm. in diameter. These are spectacular plants.

There is a strict form that has been given the cultivar name of *Tillandsia ionantha* 'Apretado.'* Instead of the usual *T. ionantha* morphology of slightly spreading leaves, especially at the tips, the blades of *T. ionantha* 'Apretado' are tightly constricted. This cultivar is quite rare.

Another cultivar of *Tillandsia ionantha* has been labeled *T. ionantha* 'Peanut' by those in the southern U.S. who originally discovered this small form. An adult does not often grow more than 4 cm. high. It is a desirable little plant that soon forms an attractive clump.

*First use of this name.

A.

B.

C.

A. *Tillandsia ionantha* v. *scaposa.*

B. & C. Front and rear views of a clump of caulescent *Tillandsia ionantha* v. *vanhyningii.* Notice the stem in rear view.

D. The interesting and rare *Tillandsia ionantha* 'Apretado.'

E. *Tillandsia ionanatha* 'Rosita.' Two groups of two plants. On left is a two-plant clump from wild. The smaller plant is ancestral. Offset of right clump was grown in cultivation. Note size and color difference between the two offsets.

Tillandsia ixioides

Grisebach; 1879

(ix-ee-oh-EE-des)

subgenus: *Anoplophytum*

***Ixioides* is a Greek combination of *ixi-*, which is derived from *Ixia*, a genus in the plant family *Iridaceae*, and the Greek suffix, *oides*, which means "like" or "resembling." A number of species in the genus *Ixia* have yellow, spreading flowers as does *Tillandsia ixioides*.**

Tillandsia ixioides is an epiphyte that grows from near sea level to almost 2,400 meters. It is widely distributed, growing in mostly dry, woody locales in Bolivia, Paraguay, Uruguay, and Argentina. *Tillandsia ixioides* sometimes develops into clumps of hundreds of plants the size of which can be over a meter in diameter! These huge balls are frequently home for sundry colonies or hives of large man-eating ants, wasps, or bees, hence the clumps must be approached *con mucho cuidado*.

Tillandsia ixioides is a medium size species, growing 15-23 cm. across and about 14 cm. in height at maturity. The plant forms a spreading rosette. The polystichous leaves are a silvery gray-green that have an almost bluish cast. They are covered with appressed, cinereous scales. The blades are narrowly triangular, stiff, and brittle so that collecting and transporting *T. ixioides* is invariably a problem because leaves are easily damaged. Consequently, many available plants have some leaf damage.

The color of the simple inflorescence and sublepidote, imbricate floral bracts is stramineous (straw-colored). The attractive, polystichous flowers are bright yellow, an uncommon color in *Tillandsia*.

Tillandsia ixioides adapts well to a variety of care conditions, but as with all species, one should keep in mind that light and water levels must be complementary. Because *T. ixioides* is xeric, it grows more rapidly with high light levels.

Tillandsia ixioides is more prone than most other species to that small, fluffy white insect, the mealybug. Normally, little if any real damage is done to the plant unless the infestation is severe. The insects usually cluster around the plant base, especially in clumps of plants. For plants belonging to the fastidious collector, or for plants to be entered in a competitive show, ANY creatures are *verboten*. Mealybugs can be removed with fingers, alcohol, or a pesticide.

Besides a superficial resemblance to *Tillandsia concolor*,* *T. ixioides* is sometimes confused with *T. xiphioides* because the pronunciation of their names is similar. They are easy to differentiate if one remembers that *T. ixioides* has more brittle, thinner, silver-green leaves and yellow flowers, while *T. xiphioides* has more flexible, broader, brownish-green leaves and large, white flowers.

**See* description of *Tillandsia concolor*, pg. 38.

A.

B.

C.

Tillandsia juncea

Poiret; 1817

(JUNE-say-a)

subgenus: *Tillandsia*

***Junceus* is an adjective that means "pleasant," "agreeable," or "delightful."**

Tillandsia juncea is widely distributed, growing from Mexico and the greater Antilles to Bolivia. Most available plants are from Guatemala and Mexico, and the morphology of *T. juncea* from these two countries, unlike a number of other species, does not differ much. However, an uncommon form from Ecuador is quite different.

Tillandsia juncea is a mainstay, found in almost everyone's collection. This is because it is readily available, inexpensive, interesting to look at, and hardy if placed outdoors during the growing season.

The Central American form is stemless and about 25 cm. across. Height varies greatly, but an average plant is 25-40 cm. Specimen plants of 60 cm. are not uncommon. The lepidote sheaths are ferruginous. The filiform blades have an indumentum of subappressed trichomes and are usually grayish-blue. Sometimes the blades have a colorful carmine cast.

The tall, slender scape and primary bracts are often carmine, and the inflorescence is usually densely digitate. The spikes are lanceolate and slightly complanate. The distichous, imbricate, and lepidote floral bracts vary from carmine to fern green. The flowers are royal purple.

The Ecuadorian plant, also stemless, is usually smaller than the Mexican-Guatemalan form, generally growing no more than 25 cm. in height. The leaves are combinations of bright green, yellow-green, almost yellow, and various shades of burgundy and orange. It is stoloniferous, which means that it produces a runner from the base that is covered with cataphyllic (small, scaly) leaves.

In Guatemala, *Tillandsia juncea* grows epiphytically in cloud forests that receive substantial precipitation for part of the year. The plants normally grow in large, dense clumps. The filiform leaves are an indication of significant air movement in the native habitat.

As with many species, such as *Tillandsia fasciculata, T. juncea* roots readily with many thin, holdfast roots. When grown in a loose, fast-draining mix, the roots of the plant will soon fill the pot. The late David Barry noted that the great expansion of roots in pot-bound plants indicates *T. juncea* can absorb water and nutrients through the root system.*

Tillandsia juncea grows well with average *Tillandsia* care along with plenty of air circulation. It is not the easiest species to grow indoors, preferring the rigors of outdoor life. Also, *T. juncea* is a relatively slow growing plant. The pups usually take two or even three years to reach maturity. The tall, slender bloom spike develops for almost a year before flowering commences.

This species flourishes when mounted in any position, even upside down. Future pups will grow lightward until the plants crowd each other and begin surrounding the mounting material.

*Barry, Jr. D. 1953. Hold-Fast Roots. *The Bromeliad Society Bulletin*. Vol. 3. No. 5. Los Angeles, Calif. pg. 45.

A.

B.

C.

A. & B. Guatemalan form of *Tillandsia juncea.*

C. Ecuadorian form of *Tillandsia juncea.*

Tillandsia latifolia

Meyen; 1835

(lah-ti-FO-lee-a)

subgenus: *Allardtia*

Latifolia **is a combination of the adjectival prefix *lati-*, which means "broad" or "wide," and the adjectival ending *-folius*, which means "leaved."**

Tillandsia latifolia is a species that grows terrestrially, saxicolously, and epiphytically in Peru and Ecuador at elevations from sea level to 2,900 meters. It is most abundant growing root-anchored in sand along the coastal desert of Peru.

Tillandsia latifolia is extremely variable, and it is often hard to believe that some of these forms are not actually distinct species. Some are viviparous (offset from the apex of the inflorescence) and some are not. Some are very large, growing to over 45 cm. in width and a meter in length, while others are often only 5-8 cm. in width and height. New forms of *T. latifolia* may still be awaiting discovery.

Most coastal desert varieties of *Tillandsia latifolia* produce viviparous offsets, although some forms produce numerous basitonal pups before blooming, so a beautiful clump develops rapidly. The weight of the developing pup eventually causes the old, decaying bloom scape to lean over far enough for the pup to make contact with the coarse sand. The new plant begins to develop a prodigous root system that penetrates the sand and anchors it. One often sees three or four generations of plants alongside one another, connected by the sclerified (woody) stem of the previous generation.

Tillandsia latifolia v. *latifolia* is widely distributed, growing in great stands in many locations along the Peruvian coast. Mature plants are caulescent and average 17-25 cm. in diameter and 15-30 cm. in length. The spreading leaves are triangular-attenuate and relatively broad when compared to some of the other forms. The fern green leaves are densely lepidote, which gives them a grayish-green cast. The plant is viviparous, but also produces offsets among the leaf axils.

The scape, primary, and floral bracts of the lepidote, compound inflorescence vary from copper to dark cherry among different plants. The small corollas of all *Tillandsia latifolia* varieties are rose to purple.

At present, there are two forms of *Tillandsia latifolia* 'Enano.' They are probably the smallest forms of this species, not usually exceeding 8 cm. in diameter and 6 cm. in height in the natural habitat.

The blades of one form are only slightly more than 1 cm. in diameter. It grows in an area of approximately two acres along the coast north of Lima. *Tillandsia latifolia* 'Enano' is caulescent but to a lesser degree than *T. latifolia* v. *latifolia*. The leaves of this plant are quite succulent. *Tillandsia latifolia* 'Enano' also produces an erect, lepidote bloom scape that is copper to dark cherry. The plant is viviparous in addition to producing basitonal offsets.

There is a slightly different form of *Tillandsia latifolia* 'Enano.' It is similar in size and appearance to *T. latifolia* 'Enano,' but it is found in a different geographical location. The blades are broader—often 2 cm. in diameter. Also, the plant is browner and lighter in color.

Tillandsia latifolia 'Delgado'* grows two or three kilometers inland from *Tillandsia latifolia* 'Enano.' The morphology is much larger than *T. latifolia* 'Enano'—40-60 cm. in diameter and 20-25 cm. high. The leaves are long and slender. Normally, the trichomes are densely distributed, which causes the plant to appear gray. The tall, slender inflorescence is the same copper to cherry as *T. latifolia* 'Enano.' The scape is taller and thinner.

Tillandsia latifolia 'Graffiti' is a large, broad-leaved plant that grows south of Lima. The stark, bleak landscape is barren except for the few patches where this form of *T. latifolia* flourishes in gullies and at the base of hills. Intrepid individuals have ingeniously transplanted a number of specimens up to the sides of promontories and spelled out their names or messages for posterity.

An adult specimen of *Tillandsia latifolia* 'Graffiti'*may be about 40-60 cm. in diameter, which is much larger that *T. latifolia* v. *latifolia*. The broad, succulent leaves are long, attenuate, and slightly channeled at the base. Although the plant has a substantial covering of appressed trichomes, the dominant color of the leaf is green. This cultivar offsets from the leaf axils and is not viviparous. It usually has a one-year growing cycle.

A.

B.

C.

A. Specimen of *Tillandsia latifolia* v. *latifolia*.

B. & C. *Tillandsia latifolia* 'Graffiti.'

D.

E.

D. *Tillandsia latifolia* 'Delgado' growing in the Peruvian coastal desert (photo Dimmitt).

E. Rare, soft-leaved, viviparous form of *Tillandsia divericata*.

The beautiful, compound inflorescence consists of a short scape with inflated, distichous orange spikes. The imbricate floral bracts are visibly dotted with white scales. *Tillandsia latifolia* 'Graffiti' does not ship well. Being quite large with stiff and brittle leaves, the plant is hard to transport without damaging the foliage.

Tillandsia latifolia v. *divericata* is a large, caulescent form that matures when 35-100 cm. in diameter. It is sometimes found hanging from cliffs with the shoot tip recurving upward. Other populations grow epiphytically in dry inland locales in Peru and Ecuador. The leaves are sometimes softer, more silver, and not as succulent as other forms of *T. latifolia*. This form varies greatly in size over its broad distribution in Ecuador and Peru. *Tillandsia latifolia* v. *divericata* has a tall, slender inflorescence with lateral to spreading bloom spikes. The distichous, linear-lanceolate spikes are complanate and cherry to cardinal. The glabrous floral bracts are imbricate. The flower petals form an interesting windmill pattern—instead of forming a tight tube, one margin of each petal overlaps the margin of the next petal and the rest of each petal is free.

Tillandsia latifolia v. *divericata* 'Gigante'* grows to be a very large plant, attaining a width of 80 cm. and a height of 100 cm. when in bloom. The cinereous, densely lepidote leaves are relatively rigid. The blades are suberect to spreading, the tips recurved. The bloom is similar to *T. latifolia* v. *divericata*, but the plant is viviparous, producing an offset from the apex of the bloom.

All of these forms of *Tillandsia latifolia* are interesting plants, easy to care for, tolerant of neglect, and each is beautiful in its own way.

*First use of this name. New Latin descriptions for the many subspecies of *Tillandsia latifolia* will be published in the future, and the new, botanically correct epithets will supercede these names.

F.

G.

H.

F. *Tillandsia latifolia* 'Delgado.'
G. *Tillandsia latifolia* 'Enano.'
H. *Tillandsia latifolia* v. *divericata.*

Tillandsia lorentziana

Grisebach; 1874

(lore-entz-ee-AH-na)

subgenus: *Anoplophytum*

This species is named after the German botanist Dr. Paul Lorentz (1835-1881).

Tillandsia lorentziana grows epiphytically or saxicolously at elevations of 700-2,600 meters in Bolivia, Brazil, Paraguay, and Argentina. One interesting habitat is in northwestern Argentina.

The terrain is mountainous and rocky. Vegetative growth consists mainly of cacti, acacias, sundry shrubs, and bushes. It is exciting to see an arroyo playing host to stands of large cacti, bushes, and trees, each having its own predominant clumps of an interesting *Tillandsia*. Although there is no real drought season and the humidity is usually high, the solar radiation is intense. Because of this, *T. lorentziana* qualifies as a true xerophyte.

Tillandsia lorentziana can be found growing along with *T. duratii, T. ixioides, T. muhrii, T. myosura* Grisebach ex Baker, *T. reichenbachii*, and numerous miniature species such as *T. bryoides* Grisebach ex Baker, *T. tricholepis* Baker, and *T. capillaris* Ruiz & Pavon.

Tillandsia lorentziana produces many offsets, and after a few years, the shoots develop into a large grouping. In nature, clumps are often home to many types of insect as are species such as *T. ixioides*.

B.

C.

The plant has a short stem and polystichous leaves. An average mature plant is about 20-25 cm. in width and height. The leaves are erect at the base and spread slightly from the midpoint to the apex. They are narrowly triangular, attenuate, and covered with spreading, cinereous scales. The foliage of plants in some populations is greener and more glabrous than others. With good growing conditions, this species soon develops a substantial root system.

Tillandsia lorentziana is a hardy plant that produces a large, cherry inflorescence. The scape is slender and erect. The inflorescence is usually bipinnate but sometimes it is simple. The spikes are oblong-lanceolate and complanate. The distichous, glabrous to subglabrous floral bracts are imbricate and the small flower petals are white.

As with other xerophytes, *Tillandsia lorentziana* requires bright light to thrive. Growing seasons of warm temperatures, frequent watering, and a generous fertilization program will produce large, strong plants.

A.

B. & C. *Tillandsia lorentziana* growing epiphytically on cactus in Andean foothills of Argentina.

Tillandsia meridionalis

Baker; 1888

(meh-ri-dee-oh-NAHL-is)

subgenus: *Anoplophytum*

***Meridionalis* is an adjective that means "south" or "southern." The species was probably named for its southern latitude range.**

Tillandsia meridionalis is an epiphyte that abounds in Argentina, Brazil, Paraguay, and Uruguay at altitudes from near sea level to 2,200 meters. It is hardy and grows well under a myriad of conditions.

A typical *Tillandsia meridionalis* habitat is the Great Chaco region of Paraguay and Argentina. Its distribution covers many hundreds of square miles. This area receives precipation only two or three months a year, but torrential rains submerge the ground in large areas for weeks at a time.

The terrain is populated by scattered scrub bushes and trees. *Tillandsia meridionalis* grows in this arid habitat along with *T. vernicosa* Baker and the more noticeable *T. duratii*. *Tillandsia meridionalis* tends to grow singly or in small groups. It does not

C.

often attain the large clump dimensions of a species such as *T. ixioides*.

An average plant is about 12-15 cm. in diameter and height. *Tillandsia meridionalis* forms a stemless rosette and the polystichous leaves are frequently secund, that is spreading to one side. The leaves are narrowly triangular and attenuate. They are prominently nerved with an indumentum of appressed, cinereous trichomes.

In addition to its durabililty, *Tillandsia meridionalis* has achieved great desirability among collectors because of its large, brilliantly colored bloom. The scape is slender and long enough to exsert the simple inflorescence above the leaves. The imbricate, polystichous, and subglabrous floral bracts are bright carmine. Gradually the bracts open out and back while the white corollas emerge. As with many of the species in subgenus *Anoplophytum*, the three petals of the corolla form a small tube with flared tips.

Although *Tillandsia meridionalis* does not root as quickly as some other species, it does eventually attach itself to the growing substrate.

The stiffness and grayness of the leaves indicate that *Tillandsia meridionalis* grows well in brightly lit areas. This in turn warrants more frequent watering to maximize growth in hot, dry locations.

A.

B.

C. Adjusting to the rigors of Chaco life in Paraguay, Jerry Robinson begins his morning shave.

Tillandsia muhrii

Rauh; 1986

(MOOR-ee-ee)

Subgenus: *Anoplophytum*

***Tillandsia muhrii** was named for an Argentine, Dorothea Muhr, who sent the plant to Professor Rauh in 1973.**

Tillandsia muhrii is endemic to northwestern Argentina at elevations of 1,200-1,600 meters. In 1983, Jerry Robinson and I found this species growing on the sides of rock outcroppings. To reach the plants required a bit of rock climbing. The find was exciting because a few of the plants were in bloom and these were quite unlike anything previously encountered. However, as thrilling as the possible discovery of a new species was, I also paid my dues. At another location an hour before, a sandstone ledge I was standing on gave way and I landed heavily on my side three meters below. For the next few weeks any physical exertion reminded me in no uncertain terms of my own vulnerability.

The narrowly attenuate, prominently nerved (the closely and evenly spaced parallel lines running longitudinally along the length of the leaf) epidermis of *Tillandsia muhrii* is similar to those of *T. xiphioides, T. diaguitensis,* and *T. peiranoi*. The silver-brown leaf color is also similar to these other species, but *T. muhrii*, like *T. diaguitensis* and *T. peiranoi*, tends to be more silver than *T. xiphioides*.

C.

Mature plants are about 10 cm. wide and 12 cm. tall. *Tillandsia muhrii* does not have as many leaves nor are they as succulent as those of *T. xiphioides*. The leaf margins of this species often have large, ciliate trichomes unlike *T. xiphioides,* which has foliar trichomes on the margins.

The simple, distichous, and complanate inflorescence is strikingly beautiful. The imbricate, glabrous floral bracts are blood red. The red bracts contrast dramatically with the silvery leaves and flared, indigo petals. *Tillandsia muhrii* is an exquisite species even though the blooms are not fragrant.

In nature, this plant often grows in sunny locations. With good air circulation, bright light, and frequent watering along with a sensible fertilization program, *Tillandsia muhrii* will grow rapidly. These conditions will also enhance overall plant size, strength, bloom, and reproductive capability.

*This species was known as *T. xiphioides* v. "unknown" for the first two years after its introduction in 1982. Subsequently it was referred to as *T. friesii*.

A.

B.

C. *Tillandsia muhrii* growing as a saxicole in Argentina.

Tillandsia paleacea

Presl; 1827

(pah-lay-AH-say-a)

subgenus: *Phytarrhiza*

***Paleatus* is an adjective that means "chaffy" or "chafflike." The name refers to the coarse, chafflike tomentum (woollike covering) of trichomes on the leaves.**

Tillandsia paleacea is found from southwestern Colombia to Bolivia and in northern Chile at elevations from sea level to over 3,000 meters. It is a versatile species that grows on cliffs and rock outcroppings, epiphytically in dry habitats, and on sand along the Peruvian coastal desert.

Different populations vary greatly in size. In cultivation, the small variety found along the Peruvian coast is only 6 cm. wide and 8 cm. long. Some of the cliff dwelling specimens grow to be as much as 20 cm. wide and 35 cm. long.

The coastal species is especially interesting because of its environmental adaptation. *Tillandsia paleacea* proliferates in large stands, each of which may cover many acres. Within these habitat sites the plants grow in small mounds about a meter across. The mounds are always oriented with the live growing tips of the shoots directed toward the fog that comes in from the coast. The plants grow for many years and achieve considerable length, but normally only 5-8 cm. of new growth is alive. The mounds form because the clumps block some of the sand blowing inland as a result of the frequent breezes off the ocean.

This caulescent species has polystichous leaves that are contorted and few-ranked (widely separated from each other). The narrow, involute-subulate blades are densely covered with a tomentum of coarse, cinereous trichomes. The older leaves are fuscous (dusky brown). In the natural environment, dead leaf and stem material in the coastal variety becomes almost black, and stems of the individual plants sometimes measure up to 60 cm. in length.

The slender scape can be short or long and is mostly glabrous. The inflorescence is simple or compound, and the spikes are lanceolate and complanate. The distichous, imbricate, floral bracts are subglabrous to glabrous. All the forms of *Tillandsia paleacea* have lovely, fragrant, aster violet blooms. The offsets are acrotonal, which allows succeeding generations to grow along the same stem.

Because of the limited available nutrients and moisture during much of the year, this species is highly oligotropic, which means that it has evolved to cope with a nutrient deficient environment. Because of these harsh conditions, one trade-off for survival in nature has been a slow rate of growth. However, *Tillandsia paleacea* does respond with faster growth to a vigorous light, water, and fertilization program.

Since the leaves are thin, the plant needs humidity and/or frequent watering. Because of the dense tomentum, it prefers a brightly lit location. The plant also does not produce many roots, so it is more likely to break loose from its mounting as the basal leaves die back. If this occurs the plant can easily be reattached.

A.

B.

C.

A. Large saxicolous variety of *Tillandsia paleacea* in bloom.

B. Coarse foliar trichome tomentum of *Tillandsia paleacea*.

C. Flower enlarged approximately two times to show more detail.

Tillandsia paucifolia

Baker; 1878

(pow-si-FO-lee-ah)

subgenus: *Tillandsia*

***Paucifolius* is a combination of the noun *pauci*, which means "few," and the adjective *-folius*, which means "leaved." This species is relatively few-leaved.**

*Tillandsisa paucifolia** is an epiphyte widespread in Florida as well as the Caribbean, Central America, Colombia and Venezuela. It grows at elevations from near sea level to 1,500 meters.

Tillandsia paucifolia is a myrmecophyte, but the base is only slightly inflated. Like *T. caput-medusae*, the sheath margins are often beet purple. They are broadly ovate and merge gradually with the blades. The involute-subulate blades are more succulent, stiffer, and straighter than other species in this group. An average mature specimen of *T. paucifolia* is about 15 cm. wide and 18 cm. in height.

Tillandsia paucifolia is normally lettuce green. It has an indumentum of fine, closely appressed, cinereous trichomes. When in bloom, the upper portion of the plant often becomes erubescent (rosy), which enhances the carmine color of the inflorescence.

The inflorescence can be simple but it is usually compound. The bloom spikes are suberect to erect, complanate, and linear-lanceolate. The floral bracts are distichous, densely imbricate, and lepidote. The flowers are mauve.

Tillandsia paucifolia grows slowly, often needing two growing seasons to reach maturity. It produces a substantial root system and responds to good cultivation. If one has patience, a large clump can be produced in a few years.

Until recently, this species was known as *Tillandsia circinnata*. But in 1982, Wilhelm Weber of East Germany published an article in the *Journal of the Bromeliad Society* that brought something very interesting to light. He showed in logical detail, supported by photographs, that the type specimen of *T. circinnata* was actually the same as *T. streptophylla*—which meant that *T. circinnatta* was not valid. The next earliest valid epithet for this particular plant, *T. paucifolia*, had been originally published by Baker in 1878.

There are Mexican populations of *Tillandsia paucifolia* 'Dimmitt's Delight'** that are viviparous. The plants not only produce basitonal offsets, but also develop one or more pups from the inflorescences. Consequently, the shoots have the ability to branch and clump into long chains of individual plants, all of which are attached to each other by the spent, sclerified inflorescences of previous generations. The viviparous form of *T. paucifolia* grows significantly larger and has a somewhat different morphology than the non-viviparous form. The blades of the former are much longer, spreading, and recurving at the tips. An average adult is 30 cm. across and 25 cm. in height. *Tillandsia paucifolia* 'Dimmitt's Delight' is a fascinating plant and is easy to grow. Most often, it is simply hung by a piece of wire and allowed to "do its thing."

*Supercedes *Tillandsia circinnata*.

** First use of this name. Mark Dimmitt of Tucson, Arizona grew an astonishing clump of this plant from a single specimen. Over a period of 13 years it grew to measure a meter in width and over 2 meters in height.

A.

B.

C.

B. Involute, succulent leaf with small, closely appressed trichomes (X2).

C. A clump of *Tillandsia paucifolia* 'Dimmitt's Delight.'

Tillandsia peiranoi

Castellanos; 1938

(pay-ee-RAH-no-ee)

subgenus: *Phytarrhiza*

***Tillandsia peiranoi* was named for the Argentine naturalist Abel Peirano, who first collected the plant.**

Tillandsia peiranoi grows saxicolously in gullies and on rock outcroppings in the Andean foothills of northwestern Argentina. A caulescent species, *T. peiranoi* is usually found in dense clumps. Individual specimens of *T. peiranoi* grow 3-4 cm. in width and 4-5 cm. in height. Since the plant offsets rapidly, the overall dimension of the clump will increase progressively, especially after the first generations.

The polystichous leaves have an indumentum of cinereous, subappressed scales. The erect to recurving blades are narrowly triangular. The appearance of the epidermis is strikingly similar to those of *Tillandsia xiphioides, T. diaguitensis,* and *T. muhrii*, species with which it shares a general habitat range. These species all have the same basic leaf color, texture, and veining. However, in size and overall plant configuration, *T. peiranoi* is quite different; specimens are much smaller and have densely polystichous, more succulent leaves.

C.

The similarity of some epidermal characteristics among these species is an example of convergent evolution for that feature. All the other species are in subgenus *Anoplophytum*. This means there are significant differences in floral structure between *Tillandsia peiranoi* and the other species.*

The beautiful, dark maroon inflorescence is simple or occasionally has one or two small branches. The scape is slender, flexuous, and subglabrous. The glabrous, distichous floral bracts are erect and almost twice as long as the internodes (space between successive bracts). The size of the inflorescence and number of floral appendages vary in number from plant to plant. *Tillandsia peiranoi*, with pale blue flowers, is one of the smallest scented *Tillandsia* species.

Tillandsia peiranoi will not dehydrate as quickly as many other small species, such as *T. albertiana*, because the leaves are more succulent. But since this species grows in dense clumps, care should be taken that the leaf surfaces have ample opportunity to dry between waterings. *Tillandsia peiranoi* can tolerate strong light levels and is a hardy species. The plant is often grown best in a well-lit patio where it has access to fresh air movement.

**See* section on Taxonomic Division of Genus *Tillandsia Subgenera*, pg. 176.

A.

B.

Tillandsia plumosa

Baker; 1888

(ploo-MOE-sa)

subgenus: *Allardtia*

***Plumosus* is an adjective that means "feathery," a reference to the coarse trichomes that give this species a fuzzy appearance. *Plumosus* is derived from the noun *pluma*, the Latin word for a "small, soft feather."**

Tillandsia plumosa grows as a saxicole or epiphyte in central Mexico, occurring at altitudes of 1,600-2,600 meters. It is one of the more attractive tillandsias because of its symmetry, hemispherical shape, and generous covering of coarse trichomes. As one would surmise from the tomentum, this plant grows in xeric habitats exposed to high levels of sunlight. *Tillandsia plumosa*, more than any other species, looks like a creature of the sea—a snow-covered sea urchin.

Tillandsia plumosa is stemless with tomentose, filiform-subulate blades. The elongated, linear trichome wings that are almost perpendicular to the leaf surface are also an indication that this plant receives moisture (fog) frequently at certain times of the year. Coarse scales allow the epidermis to dry more quickly. This helps prevent possible suffocation by permitting gas exchange to take place sooner than if the scales had more spreading or appressed wings.*

Tillandsia plumosa is similar in appearance to *T. ignesiae* Mez and *T. atroviridipetala*.** A mature specimen of *T. plumosa* is 14-18 cm. in diameter, which is larger than either *T. ignesiae* or *T. atroviridipetala*. Although there is a suite of character traits that separate these two species, one easy method to differentiate them is to observe the bloom. If the inflorescence is densely capitate, the plant is *T. plumosa*. If the inflorescence is simple, the plant is *T. ignesiae*.

Tillandsia plumosa is often confused with *T. tectorum* because the two species superficially resemble each other. However, they are easily distinguishable because *T. plumosa* is hemispherical in shape while *T. tectorum* is usually more rectangular (although *T. tectorum*, too, can be hemispherical in rare cases). Another differentiation between the two is that *T. tectorum* is normally fuzzier and whiter than *T. plumosa*. This is because the trichome wings of *T. tectorum* are larger and more dense. Of course the species' blooms are unalike.

The bloom of *Tillandsia plumosa* is strikingly colorful. The slender scape and lanceolate, complanate spikes are carmine. The distichous, imbricate floral bracts are lepidote—which bestows an attractive, frosty look to them. The small petals are fern green which is unusual for a *Tillandsia*.

Tillandsia plumosa usually produces two or three pups that grow up among the leaves of the parent plant. As the pups develop, they gradually merge with and replace the parent.

Tillandsia plumosa should be cultivated in the manner of other xeric species, but it should be mounted with the plant base perpendicular to gravity. When mounted in this manner, water does not have a chance to wet the meristem for long periods, possibly causing it to rot.

**See* subsection on Trichome Variations, pg. 208, for more discussion.

**For more discussion, refer to desription of *Tillandsia atroviridipetala*, pg. 20.

A.

B.

C.

Tillandsia polystachia

Linnaeus; 1762
(pol-i-STAH-key-ah)
subgenus: *Tillandsia*

***Polystachius* is of Greek origin and is a combination of *poly-* and *-stachys*. *Poly-* is a prefix that means "much" or "many" and *-stachys*, which originally meant an "ear of corn," is used botanically for a "spike." Therefore, *polystachia* means "with many spikes," a reference to the number of bloom spikes.**

Tillandsia polystachia, one of the earliest described species, has a broad habitat range that extends from the West Indies and Mexico to Brazil and Bolivia. It is also one of the hardier tillandsias, as evidenced by its extremely diverse habitat distribution. The plant grows as an epiphyte from near sea level to more than 1,800 meters.

Tillandsia polystachia is a stemless species that varies greatly in size, but an average mature plant is approximately 30 cm. in width and 25 cm. in height. The polystichous, usually flat, narrowly triangular leaves form a tank rosette that holds a considerable quantity of water. As with any tank bromeliad, water kept in the cup of the plant for any length of time should be kept fresh. The leaves are arched-recurving and densely covered with small, appressed trichomes. Leaf margins are sometimes amaranth.

The bloom scape of *Tillandsia polystachia* is tall, slender, and erect. Although the inflorescence is rarely simple, it is usually pinnately or subdigitately compound. The spikes are erect, linear, and complanate. They vary in number and are usually dense. The distichous, imbricate, glabrous to subglabrous floral bracts are most often a combination of green and blood red to amaranth. The flowers are grape violet. *Tillandsia polystachia* normally produces numerous pups that mature in a year if well grown.

Tillandsia polystachia is an adaptable species and flourishes with either high or low light and/or water levels. One should make sure that the amount of water and light are commensurate.

Low light cultivation produces dark green foliage. With high levels of light, the leaves become yellow green, sometimes with a rosy cast.

Easy to cultivate, *Tillandsia polystachia* grows well under many environmental conditions. However, the leaf axils and root base are sometimes prone to mealybugs. The insects are seldom numerous and are rarely harmful to the plant. They are easily removed with rubbing alcohol or any mild pesticide. *Tillandsia polystachia* can be grown terrestrially in a loose potting mix, in which case it will develop an extensive root system.

Tillandsia pseudobaileyi

(Rose ex Small) Gardner; 1984*

(sue-doe-BAY-lee-ee)

subgenus: *Tillandsia*

***Pseudo-* is from the Greek and means "false," that is resembling but not equaling. This name was given to the species because for many years it was confused with *Tillandsia baileyi*, another species that was originally collected in Texas by V. Bailey in 1900.**

Tillandsia pseudobaileyi occurs from Mexico to Nicaragua. *Tillandsia baileyi* is endemic to southern Texas and Tamaulipas, Mexico. *Tillandsia pseudobaileyi* has a much broader habitat range and is much more common in cultivation than *T. baileyi*.

Tillandsia baileyi is usually smaller than *T. pseudobaileyi*. *Tillandsia baileyi* averages about 15 cm. across and 20 cm. high, while *T. pseudobaileyi* is closer to 20 cm. in width and 30 cm. in height. The inflorescence of *T. bailey* is usually simple, while that of *T. pseudobaileyi* is most often compound.

Tillandsia pseudobaileyi is a myrmecophyte that is easy to recognize because the inflated base is similar in appearance to an onion. Besides green, the basic color is amaranth, but since most of the plant is densely covered with finely appressed trichomes, *T. pseudobaileyi* also has a frosted, silvery appearance. The leaves are striated longitudinally with thin amaranth lines. Overall, it is a striking species, albeit one has to look closely because the colors and markings are subtle.

C.

The stiff leaf blades are involute, subulate, and often upwardly curved and contorted along their lengths. Because of their shape and rigidity, they sometimes appear as an interesting, modern abstraction.

The inflorescence is similar in shape to that of *Tillandsia caput-medusae, T. butzii,* and *T. bulbosa*. The color, however, is more reflective of the amaranth color of the plant, and the scape is relatively tall and erect. The individual spikes are lanceolate, and the floral bracts are densely imbricate. The flowers are indigo. *Tillandsia pseudobaileyi* produces offsets and roots readily, although the plant is relatively slow growing, the pups often needing two years to mature.

Tillandsia pseudobaileyi flourishes in brightly lit outdoor patio or garden situations but needs frequent watering in dry climates and a standard fertilization program spring through fall. *Tillandsia pseudobaileyi* is an ageotropic species, so it grows well horizontally or even upside down, as do other myrmecophytic species.

*Gardner C.S. 1984. New species and nomenclatural changes in Mexican *Tillandsia*. Sarasota, Florida, Selbyana 7: 363.

A.

B.

B. Detail showing greatly involuted—even overlapping—margins of healthy leaves.

C. Blooming specimen of *Tillandsia baileyi.*

Tillandsia punctulata

Schlechtendal & Chamisso; 1831

(punk-too-LAH-ta)

subgenus: *Tillandsia*

***Punctulatus* is an adjective that means "minutely dotted." The suffix *-ata* indicates possession. *Punctulatus* is a reference to the fine spotting on the floral tube.**

Tillandsia punctulata grows epiphytically in forests at elevations of 350-2,000 meters in Mexico and Central America. A popular species, this striking plant has bright green blades and dark purple sheaths that appear almost black. *Tillandsia punctulata* forms a rosette that is at once symmetrical, graceful, and reminiscent of a small pineapple plant. Numerous stoloniferous pups are often present.

The leaves form a dense, stemless rosette. The slightly channeled or flat blades are long and attenuate. The bases are broadly ovate.

The inflorescence is simple and erect, lanceolate, and inflated. The floral bracts are distichous. In nature, the scape and primary bracts are often brilliant scarlet. The floral bracts are green, and the flowers are indigo and white-tipped.

Tillandsia punctulata often grows deep in the forest canopy. The dark indigo of the sheaths is actually a pigment, anthocyanin. Benzing has theorized that the dark color of the sheaths may possibly serve one or both of two functions.* Anthocyanin reflects light from the red portion of the spectrum back into the green of the blades, where it is absorbed by chlorophyll. Light reaching the plants deep in the canopy would be red-enriched (of the longer wavelengths). The chlorophyll uses light to produce photosynthate, which is in turn converted into useful energy.** The other possibility for the dark coloration is that the sheaths may aid cryptic animals (certain types of larvae, etc., that like to hide in dark places) that live in the leaf axil chambers in symbiosis with the plant. The creatures are provided shelter, water, and possibly food, while the plant receives mineral and nutrient benefits.

An important consideration with this species, and any other that holds water for extended periods among the leaf axils, is to make certain that the water remains fresh and does not become stagnant. This is easily accomplished with frequent watering (which should be performed anyway). Plant rot due to the accumulation and retention of stagnant water often causes the center of the plant to suddenly come out. In addition, a putrid odor permeates the air.

Tillandsia punctulata does not require bright light outdoors, even though the plant grows surprisingly well with good light (if it receives a compensatory amount of moisture). *Tillandsia punctulata* is a mesophyte and prefers a cool environment. In its natural surroundings, the plant receives plentiful and frequent rainfall. *Tillandsia punctulata* enjoys high humidity because it has relatively few trichomes to act as barriers to leaf transpiration. It will dehydrate quickly if cultivated in an arid environment.

If one purchases or receives a bare-root specimen, the plant should be mounted with a swath of sphagnum moss around the root base to approximate its natural condition. Only a few species, such as *Tillandsia punctulata*, *T. cyanea*, and the green-leaved, water-impounding, rosette species, prefer large amounts of moss and/or soil for the root system. In the wild, the roots of *T. punctulata* play a significant role in the absorption of moisture and nutrients.

*Personal communication.

***See* section on Photosynthesis, pg. 190.

A.

B.

B. Note the small stoloniferous offsets at the base of the plant.

Tillandsia purpurea

Ruiz & Pavon; 1802
(poor-POOR-ay-a)
subgenus: *Phytarrhiza*

***Purpureus** means "purple," that is a "dull red with a slight dash of blue." The name is a reference to the color of the bloom spikes and floral bracts.**

Tillandsia purpurea proliferates in great stands as a ground cover along Peru's coastal desert. Like *T. paleacea* and *T. latifolia*, with which it shares this stark environment, *T. purpurea* grows only in localities where fog sufficiently saturates the leaves with water in the wet months from July to October. The weather in November, May, and June is variable. From December through April, the plants rarely receive water.**

Tillandsia purpurea is an exquisite plant with heavily fragrant blooms. The leaves are soft and velvety, although quite rigid and unforgiving if bent. Because they break so easily, the plant is difficult to ship—whether from Peru to the States, or across the city. But once established in a home, *T. purpurea* grows rapidly.

The polystichous leaves of *Tillandsia purpurea* form a

D.

spreading rosette. An evenly proportioned, symmetrical plant, it is attractive whether viewed from the side or looked down upon from above. The narrowly triangular leaf blades are covered with fine, spreading cinereous or fuscous (grayish or dull brown) trichomes.

Unlike *Tillandsia latifolia*, which has many varieties with different shapes, colors, and sizes, *T. purpurea* usually maintains a uniform look. However, isolated populations often vary greatly in size. I encountered one stand in which the plants were no bigger than 15 cm. in width and height. In another area, the plants measured as much as 35 cm.

There was one population that was stellate (star-shaped) when looked down upon. The leaves were stacked on top of each other so only the five leaf axes could be seen. In cultivation, the plants grow a long stem and produce numerous axial offsets. This plant is called *Tillandsia purpurea* 'Shooting Star.'***

Tillandsia purpurea produces a tall, erect scape with stramineous to maroon, imbricate primary bracts. The lepidote bracts are peppered with large trichomes that appear like so many grains of salt. The spikes are erect to spreading and lax. The floral bracts atop the scape are stramineous to combinations of rose pink and mauve, and lepidote to glabrous. The flared petal blades have cream centers and aster violet margins.

Tillandsia purpurea usually blooms for at least a month, subsequent to which it produces as many as a dozen acrotonal pups.

*As an historical footnote, *purpura* was originally the name for a shellfish of the eastern Mediterranean Sea that yielded to the Phoenicians a liquid for the manufacture of the celebrated purple dye of classical antiquity, later the name of the dye itself. From Stearn, W.T. 1983. *Botanical Latin*. David & Charles. North Pomfret, Vermont, pg. 492.

**Torres G., Juan and Lopes Ocana, Carlos. 1981. Productividad Primeria en las Lomas del la Costa Central del Peru. *Bulletin of Lima #14*.

***First use of this name.

A.

B.

C.

B. & C. Note density difference of trichomes on primary bracts of two plants.

D. *Tillandsia purpurea* 'Shooting Star.'

Tillandsia reichenbachii

Baker: 1889

(rye-ken-BACH-ee-ee)

subgenus: *Phytarrhiza*

Tillandsia reichenbachii **was named for the German botanist Reichenbach.**

Tillandsia reichenbachii grows epiphytically in woods from 200-2,000 meters in southern Bolivia and northern Argentina. *Tillandsia reichenbachii* looks like a tiny version of *T. duratii* but is easily distinguished from it because this species rarely grows more than 16 cm. across and 10 cm. tall.

In nature, *Tillandsia reichenbachii* is not often found growing in large clumps, and because of its size, the plant does not dominate its host as *T. duratii* sometimes does. *Tillandsia reichenbachii* is found growing singly or in groupings of a few plants, usually on thorny acacias or other equally thorny bushes. *Tillandsia myosura* is often found growing on the same shrubs.

C.

The leaves of *Tillandsia reichenbachii* are polystichous, and the narrowly triangular, involute-subulate blades are recurving and covered with an indumentum of subappressed, cinereous scales.

The scape is erect and slender. The inflorescence is compactly bipinnate and has fewer spikes than another, larger, look-alike, *Tillandsia streptocarpa*, and many fewer spikes than *Tillandsia duratii*. Also, like the other two species, *T. riechenbachii* has a delicately fragrant corolla composed of three, pale violet petals with white centers.

Because the plant is small, it is not as impressive as *Tillandsia duratii*. But *T. reichenbachii* is a good plant for those with an eye for beauty on a smaller scale or a desire for an uncommon species.

Tillandsia reichenbachii is a hardy species and grows well under the average *Tillandsia* conditions of bright light, frequent watering, and consistent fertilization during the growing season.

A.

B.

B. Note seed pods on spent inflorescences.

C. Typical Argentine terrain of *Tillandsia reichenbachii*.

Tillandsia rothii

Rauh; 1976

(ROTH-ee-ee)

subgenus: *Tillandsia*

***Tillandsia rothii* was named for an avid Southern California plant collector, Jack Roth.**

Tillandsia rothii is a beautiful, large *Tillandsia* that grows along the west coast of Mexico from near Puerto Vallarta southward. This species grows epiphytically in woods where summertime temperatures are hot and rain is rare. Consequently, *T. rothii* is difficult to find without a number of leaf blemishes. During the rainy season, *T. rothii* receives large quantities of precipitation and produces lush new growth.

Individual mature specimens of *Tillandsia rothii* vary greatly in size. An average adult is about 24 cm. wide and 15 cm. high, although large specimens can grow to be twice as large. At the same time, smaller plants may mature and bloom when only 15 cm. wide and 12 cm. high.

Tillandsia rothii grows slowly, often taking two to three years to fully mature from a pup. I once entered a specimen in a local

B.

bromeliad show when the plant was in bud, and then re-entered the same plant the next year, still in the full glory of its spectacular bloom!

In appearance, it looks like a yellow-green *Tillandsia xerographica*. The leaf blades are long, broad, attenuate, and recurving, but they do not often twist like those of *T. xerographica*. The abaxial blade surface is nerved and minutely appressed-lepidote while the adaxial surface is subglabrous.

When grown in bright light, the inflorescence of *Tillandsia rothii* is intensely colorful, large, and long lasting. As the inflorescence develops, the scape bracts, primary bracts, and even some of the leaves gradually become brilliant dark cherry. The inflorescence is compound with inflated spikes that are combinations of various shades of cherry, moss green and buttercup yellow. The imbricate floral bracts are glabrous. The flowers are a pale shade of seafoam. This species normally produces only one or two offsets.

Tillandsia rothii will often become erubescent when grown in bright light. The plant can also be cultivated in shadier locations. The foliage will be darker green and the plant will grow more slowly. It develops more quickly with bright light and frequent watering. Impounded water should not be allowed to stagnate. Gardner (1982)* contends that *T. rothii* may be an introgressant (species with a natural hybrid origin) between *T. roland-gosselinii* Mez and *T. jaliscomonticola* Matuda.**

*Gardner, S. 1982. *A Systematic Study of Tillandsia subgenus Tillandsia*. Ph.D. dissertation. College Station. Texas A&M University, pg. 153.

***See subsection on Phylogenetic Change Through Hybridization, pg. 179.*

Tillandsia schiedeana

Steudel; 1841

(shee-day-AH-na)

subgenus: *Tillandsia*

***Tillandsia schiedeana* was named in honor of C.J.W. Schiede, who collected the plant in Vera Cruz, Mexico.**

Tillandsia schiedeana is a common species found from Mexico and the West Indies to Colombia and Venezuela. In addition to the type form, *T. schiedeana v. glabrior* Gardner occurs in Mexico and was described in 1981. *Tillandsia schiedeana* 'Minor,' is an undescribed subspecies from Guatemala.

Both described varieties of *Tillandsia schiedeana* are xerophytic and usually encountered epiphytically or saxicolously in droughty environments. *Tillandsia schiedeana* 'Minor' grows epiphytically in oak forests and is more mesic.

Tillandsia schiedeana v. *glabrior* is distinguished from *T. schiedeana* by having leaves 8-10 cm. long that are generally secund. The plant is about 12 cm. in width and 15 cm. in length. The scape bracts are cherry and glabrous. The flowers vary from cherry to a combination of cherry and yellow.

The leaves are polystichous, long, narrowly attenuate, involute, and subulate. They have an indumentum of dense, cinereous trichomes that are appressed near the apex and spreading below. The trichomes of *Tillandsia schiedeana* 'Minor' are less dense than those of the other two varieties.

The leaves of *Tillandsia schiedeana* are stiffer, grayer, and thicker than those of *T. schiedeana* 'Minor.' Although the overall shape of the two forms is the same, there are significant differences in appearance. An adult specimen of *T. schiedeana* grows to about 20 cm. in width and 15 cm. in height. *Tillandsia schiedeana* 'Minor' is caulescent and grows to a length of up to 25 cm. with many lateral rows of leaves that spread up to 20 cm.

The leaves of *Tillandsia schiedeana* 'Minor' are finer, greener, and more numerous. Because of this, the plant prefers conditions somewhat more humid than its counterpart. Additionally, it will not tolerate the high light levels of *T. schiedeana*. The bloom of *T. schiedeana* 'Minor' is less colorful and smaller than *T. schiedeana*; nonetheless, its myriad of graceful leaves creates a striking plant.

When grown into clumps, leaves of individual plants intertwine. A silvery clump of *Tillandsia schiedeana* in bloom will have a plethora of simple bloom spikes that appear as many fulgent, cherry torches. The petals are vivid yellow, which is uncommon in *Tillandsia*. This is a plant that can be hung from a piece of wire or fishing line. Since it is also inexpensive, a grower can start by grouping three or four plants together in order to reduce the time necessary to grow a large clump.

For some reason, *Tillandsia schiedeana* does not seem to be a good houseplant—at least in southern California. As an outdoor species, *T. schiedeana* will thrive if given high levels of light, water, and plenty of fresh air.

A.

B.

C.

A. A clump of *Tillandsia schiedeana* 'Major,' along with some *T. usneoides*.

B. *Tillandsia schiedeana* 'Minor.'

C. *Tillandsia schiedeana* v. *glabrior*.

Tillandsia secunda

Humboldt, Bonpland, & Kunth; 1816
(seh-COON-da)
subgenus: *Allardtia*

***Secundus* is an adjective that means "turned to one side," a reference to the flowers that are all borne on the same, dorsal side of the bloom spikes.**

Tillandsia secunda is a widespread species that grows terrestrially and saxicolously at elevations of 2,000-3,500 meters in Ecuador.

Tillandsia secunda forms a rosette and at maturity is normally 30-60 cm. in diameter. The narrowly triangular leaves are prominently nerved with an indumentum of appressed, cinereous scales that give the leaves a slightly frosted appearance. The margins are an attractive amaranth that becomes more pronounced as the blades merge with the sheaths.

This species rapidly develops an elaborate system of thick, holdfast roots. When mounted on a rough piece of rock or wood, the roots soon compete with each other for space. When this occurs, they simply overlap each other, making plant attachment even more secure.

D.

The scape is erect, and the glaucous (waxy coating) primary bracts are amaranth. The spikes are spreading. The perianths are secund and face downward, a position favored by foraging hummingbirds. The flowers are grape violet.

Tillandsia secunda is one of the most interesting species because after it blooms numerous, viviparous offsets develop on the spikes of the inflorescence. Other species, such as *T. latifolia* and *T. paucifolia*, have forms that are viviparous, and a few, such as *T. somnians* L.B. Smith, produce multiple offsets along the scape. But *T. secunda* is the only one that normally produces so many offsets in this manner. As the pups grow and begin to crowd each other, they can be removed easily and attached to another medium. One often gets 15-60 new plants from a spent inflorescence.

The base of *Tillandsia secunda* can sometimes become infested with mealybugs. They can be removed with a pesticide, rubbing alcohol, etc. This species grows best with bright light, high humidity, and enough water to keep the leaves turgid. Water maintained in the cup should be kept fresh to avoid rot. Temperatures are not critical except for extreme heat and cold. When grown as a terrestrial, the plant will often grow larger than epiphytic specimens because of the role roots play in moisture and nutrient uptake. As with any broad-leaved species, one should keep the plant away from areas subject to high winds in order to avoid leaf damage.

A.

B.

C.

Tillandsia seleriana

Mez; 1903

(seh-lair-ee-AH-na)

subgenus: *Tillandsia*

***Tillandsia seleriana* was named for C. Seler, who collected the plant in Mexico and was the first to request identification.**

Tillandsia seleriana is native to pine and oak woods at elevations of 270-2,400 meters. This species is found from southern Mexico to Honduras. In appearance, it most often reminds people of an artichoke.

Tillandsia seleriana is easily distinguished from other myrmecophytic species because it is the only one in which the long, attenuate blades constrict. In virtually every instance, specimens of the other species have divergent leaves. *Tillandsia seleriana* has more space for its symbiotic tenants than any of the other ant plants.* One will often see leaves with holes chewed by ants.

This species has the largest pseudobulb—the base of the plant sometimes exceeds 15 cm. in diameter. A plant this broad at the base would be close to 30 cm. tall without a bloom spike. The average size of an adult plant is 10-15

B.

cm. across at the base and 15-20 cm. in height.

Tillandsia seleriana is stemless, with leaves that are covered with large, coarse, cinereous scales. The sheaths vary from ovate to suborbicular. Instead of a distinct delineation between sheath and blade, as with species such as *T. butzii* or *T. bulbosa*, those of *T. seleriana* merge together more gradually. The blades are deeply channeled and involute. The plant will develop an extensive root system when the base is provided with a humid environment. The inflorescence and upper blades become an attractive clear pink to carmine at blooming time. The blush of color, along with the velvety, silver trichomes, creates a soft, pastel display of color. The inflorescence is digitately or subpinnately compound. The spikes are broadly ellpitic and strongly complanate. The coarsely lepidote floral bracts are imbricate. The flowers are indigo.

Tillandsia seleriana thrives under average conditions of light, water, and fertilzer. The plant will grow well with high light levels as long as the corresponding humidity and/or watering schedule are there to support the more rapid leaf transpiration. *Tillandsia seleriana* will also adjust to lower light but must receive a corresponding decrease in watering frequency. Of course, the plant will not grow as rapidly in this type of environment. Fertilizing on a scheduled summertime basis will increase the growth rate, although *T. seleriana* is one of the slower growing species. It often needs two or three years to mature from an offset.

**See* section on Myrmecophytism, pg. 9.

A.

A. Clump of *Tillandsia seleriana* thriving on piece of tree fern. Note gradually decaying ancestral plant in middle of group.

Tillandsia sphaerocephala

Baker: 1888

(sfy-row-SEH-pa-la)

subgenus: *Allardtia*

***Sphaerocephalus* is a combination of two Greek root words. The first, *sphaero-*, means "globose" or "spherical." The second, *-cephalus*, means "headed." The "spherically-shaped head" refers to the rounded inflorescence.**

Tillandsia sphaerocephala grows as a saxicole or epiphyte at elevations of 950-3,550 meters in Bolivia and Argentina. I encountered this interesting species growing outside Sorata, Bolivia. The excursion was difficult because of the inclement weather and the steepness of the ravine beneath the cliffs where the plants abounded.

A mature specimen of *Tillandsia sphaerocephala* averages about 20 cm. in width and height. A stemless plant, *Tillandsia sphaerocephala* forms a dense rosette of narrowly triangular leaves. The sheaths are dark brown, and the gray blades have appressed, somewhat lustrous scales that cause the leaves to appear shinier than those of most other species. The blades are slightly channeled, the tips involute and subulate.

B.

The scape is 5-8 cm. tall, which is enough to exsert the rose-colored, subglobose flower head above the leaves. At first glance the inflorescence appears to be simple, but it is actually made up of a number of complanate spikes that have from one to three flowers each. The corollas are small and amaranth.

With good growing conditions, *Tillandsia sphaerocephala* rapidly produces an extensive root system. It is one of the more prolific of the entire genus in this regard. When the base is mounted close to a substrate, new roots will quickly attach the plant and grow over the mounting surface.

Tillandsia sphaerocephala is easy to cultivate under a broad range of conditions. This species is also a good houseplant if given enough light and water.

A.

B. View of Sorata, Bolivia, from across narrow valley.

Tillandsia straminea

Humboldt, Bonpland, & Kunth; 1816

(strah-MI-nay-a)

subgenus: *Phytarrhiza*

***Stramineus* is an adjective that originally meant "straw." It now means "straw-colored," and is a reference to the color of the primary bracts.**

Tillandsia straminea grows as an epiphyte in many different locations in Peru and Ecuador. The most common habitat is mountainous and xeric although stands are also found in the Peruvian coastal desert. This species often grows on cacti as well as on *Acacia* and *Erythrina* trees. While there are a number of sympatric species, *T. disticha* and *T. cacticola* are those most frequently encountered.

Mature plants vary greatly in size. Small plants often bloom when the foliage is about 23 cm. in width and height, while large specimens often grow to be 35 cm. I currently have one specimen of *T. straminea*, aka "King Stram," that measures a full 80 cm. in diameter, 60 cm. in height, and tips the scales at 1 kilo!

Tillandsia straminea is stemless and the leaves are polystichous. The narrowly triangular blades are long, attenuate, and erect. As the leaves mature, they gradually spread and begin to decurve. Old leaves deflex (hang down). The leaves of *Tillandsia straminea* are

B.

C.

longer, softer, and more finely covered with spreading, cinereous trichomes than those of the closely related species *T. purpurea* and *T. cacticola*. When it is well grown, the white leaves feel like soft velvet.

When the plant blooms it rewards the grower with a brightly colored, compound inflorescence that produces a great many fragrant flowers. The spikes are erect to spreading. The glabrous, tubular to subfusiform (almost spindle-shaped) floral bracts vary among different plants from stramineous to combinations of rose pink and mauve. The throats of the flared petal blades are cream, the margins grape violet.

Tillandsia straminea grows rapidly with favorable conditions. It flourishes in well-lit outdoor locations where it receives sufficient moisture.

Tillandsia straminea is not currently recognized by some authorities as a true species but rather as an intermediate variety of *T. purpurea*. Although the blooms of *T. straminea* are similar to those of *T. purpurea*, the scape is much taller. Also, the vegetative differences are significant and consistent; the leaves of *T. straminea* are longer, softer, and not as rigid. As with the many forms of *T. latifolia*, substantial work needs to be done in organizing the taxonomy of this group of species/varieties.

A.

B. Floral detail enlarged approximately two and one half times.

C. *Tillandsia straminea* growing on cactus in northern Peru.

Tillandsia streptocarpa

Baker; 1887

(strep-toe-CAR-pa)

subgenus: *Phytarrhiza*

***Streptocarpus* is from the Greek. It is a combination of the adjective *streptos*, which means "twisted," and the noun, *carpos*, which means "fruit." The combination is a reference to the twisted seed capsules.**

Tillandsia streptocarpa grows epiphytically in open woods or saxicolously in direct sunlight. It is found at elevations of 60-2,300 meters in Peru, Bolivia, Paraguay, and Brazil. This caulescent species appears as a smaller, thinner form of *T. duratii* and, likewise, has sweet, fragrant blooms. *Tillandsia streptocarpa* grows to be significantly larger than another species with a similar morphology, *T. reichenbachii*. An average adult is 30 cm. across and 25 cm. tall.

The blades are linear-triangular and attenuate. New leaves are erect. Mature leaves are spreading and recurve at the blade tip. Older leaves dry into ringlets in order to coil around a twig or tree branch. The blades are generally straighter and not as recurving as those of *T. duratii*. They are densely covered with spreading, cinereous trichomes. These give the plant its fuzzy white appearance.

The bloom scape is tall and slender. The inflorescence is bipinnate, that is, the bloom spikes are distichous. Then on each individual spike the imbricate floral bracts are also distichous. The broad flared petals of the flower are pale lavender and white.

The inflorescence of *Tillandsia streptocarpa* is smaller and has fewer flowers than that of *T. duratii*. However, it does bloom for nearly a month, and for those with a space problem this is an ideal, fragrant flowering species.

The complanate spikes are linear to elliptic, and arched-spreading. The imbricate floral bracts are from lepidote to glabrous.

Tillandsia streptocarpa is a xerophyte so it prefers strong light. The plant will grow faster and larger if watered frequently and fertilized in dry and/or hot climates. *Tillandsia streptocarpa* produces one or two acrotonal pups after blooming. When grown suspended from a piece of wire, a few generations will see *T. streptocarpa* develop into a tangled ball of plants that will produce a plethora of beautiful bloom spikes.

A.

B.

A. Flowers enlarged approximately two times to show detail.

Tillandsia streptophylla

Scheideweiler; 1836

(strep-toe-FIH-la)

subgenus: *Tillandsia*

***Streptophylla* is from the Greek. It is a combination of the adjective, *streptos*, which means "twisted," and *phyllon*, which means "leaf."**

Tillandsia streptophylla is a myrmecophytic species that grows epiphytically from sea level to more than 800 meters in elevation, and its range extends from southern Mexico to Honduras. This is one of the few species to have evolved a nickname—the "Shirley Temple" plant. The leaves often twist and recurve in a manner that is Nature's imitation of the coiffeured child of days gone by.

Interestingly, the grower can control the amount of curl in the leaves by the amount of water and humidity the plant receives. In a wetter, more humid environment the leaves grow straighter. With more xeric conditions, the leaves tighten into curled ringlets. Enthusiasts can be found who favor either style—it is purely a matter of personal preference as to how one wishes to grow this species. It varies in size; a small mature specimen is about 18 cm. in width and height. Larger plants often grow to 60 cm.

A plant that is dehydrated can be quickly hydrated by an overnight soaking in water—a pinch of fertilizer helps. The leaf cells will absorb water until full and then will not absorb any more. Oxygen and carbon dioxide gas exchange, so vital to a plant's metabolic activity, will not be adversely affected if interrupted for a day or so.

The sheaths of *Tillandsia streptophylla* are broadly ovate or elliptic and covered with coarse, spreading, cinereous scales. It is an exception to other species in the myrmecophytic group in that the narrowly triangular, attenuate, blades are flat instead of deeply channeled and involute-subulate.

The compound inflorescence of *Tillandsia streptophylla* is large and colorful. The scape and primary bracts are normally carmine when given bright light. The tall scape has a number of suberect to spreading, linear spikes that are complanate. The imbricate floral bracts are densely lepidote and citron green. The mauve blooms add color for weeks.

This is a species that will grow well under varied conditions. As previously stated, the care one gives the plant will directly affect its appearance. *Tillandsia streptophylla* is somewhat more susceptible to rot than many other species. This is especially true indoors where a lack of fresh, circulating air sometimes contributes to the problem of continuously moist meristem. Mounting the plant in a horizontal position often helps ameliorate this condition.

Tillandsia streptophylla is often difficult to mount because it is so bulbose. As a consequence, the base can be far from the mounting medium if the plant is mounted vertically. Even though this is not important for the health of the plant, aesthetically it is more pleasing if the roots are close to the mounting medium.*

**See* section on Mounting Tillandsias, pg. 6

A.

B.

Tillandsia stricta

Solander; 1813

(STRIC-ta)

subgenus: *Anoplophytum*

***Strictus* is an adjective that means "drawn together," "close," or "tight." With this species, the reference is either to the dense arrangement of the leaves or to the tightly imbricate floral bracts.**

Tillandsia stricta grows epiphytically in eastern South America from Venezuela to northern Argentina. Habitat conditions vary—some plants grow in mesic regions that receive frequent rainfall, while other plants endure dry periods in less mesic areas. Altitudes range from sea level to 1,700 meters.

Tillandsia stricta is a medium size species, normally 12-15 cm. in width and 12 cm. in height. But the species is variable in morphology, no doubt because of the wide range of habitats and conditions under which the plants grow.

Tillandsia stricta develops along a short stem and forms a rosette with long, narrowly triangular, attenuate, green to gray leaf blades. Sometimes the blades are spotted or margined with the dark purple pigment, anthocyanin. The coloring is normal and gives the appearance of added depth.

Tillandsia stricta is one of the most popular species because of its colorful bloom, which is approximately 4-5 cm. in length. The only characteristic that could be considered unfavorable is that the bloom is relatively short-lived. For

D.

E.

the plant to grow from bud to spent bloom often takes no more than three weeks. However, one should realize that the past-bloom plant is about to produce numerous offsets that will rapidly grow into adults. These will in turn bloom in 8-12 months.

The bloom scape extends the simple inflorescence beyond the leaves. The imbricate floral bracts vary from glabrous to appressed-lepidote and are usually brightly colored, varying from cherry to rose, although a few plants have green floral bracts. As the bloom matures, the bracts open out and spread as the light to dark blue flowers emerge.

Tillandsia stricta is a species that usually improves in appearance from the first generation to the next, as long as the succeeding plants are grown as individual specimens. Although striking, when grown as a clump the plants are smaller than if grown separately. Frequently, first generation, imported plants do not have as many leaves as plants of succeeding generations. Also, future plants are more robust, larger, more stiffly leaved, and produce larger bloom spikes.

Tillandsia stricta develops and proliferates rapidly and is easy to grow because it tolerates a wide range of cultural conditions. Truly one of the best, this species is a beautiful addition to any collection. *Tillandsia stricta* should be given bright light, sufficient water, and a consistent fertilization program.

A.

B.

C.

A & E. The most common form of *Tillandsia stricta*.

B. Closeup showing floral bracts, sepals, and petals.

C. A grayer, stiffer-leaved form that has a larger, more carmine inflorescence.

D. Specimen with enormous inflorescence.

Tillandsia tectorum

Morren; 1877

(tek-TORE-oom)

subgenus: *Allardtia*

***Tectorum* is the genative plural (the case that shows possession) of the noun *tectum*, which means "roof." For instance, *plantae tectorum* means "plants of roofs." According to Smith*, the epithet is a reference to the fact that this species has been cultivated on roofs in Peru and Ecuador.**

Tillandsia tectorum is a caulescent species that grows in large, saxicolous populations on rock outcroppings and cliff faces in Ecuador and Peru at elevations of 980-2700 meters. The large, cinereous trichomes and the interesting morphology of the plant are responsible for its great popularity. At the same time, *T. tectorum* is one of the easiest tillandsias to grow if given bright light and not over or under watered.

Tillandsia tectorum has many different varieties growing in its wide range of habitats, and mature plants vary greatly in shape and size. Smaller plants are often 5-8 cm. in width and height. Larger plants attain sizes of 25-30 cm. Rare, basketball size specimens do exist. *Tillandsia tectorum* is generally rectangular in shape. The polystichous leaves are filiform-attenuate. The phyllotaxis of plants in the Ecuadorian Andes is more dense than that of the Peruvian types.

In Peru, the plants grow at lower elevations in dry locales where cacti often dot the landscape. These plants are smaller with fewer leaves per plant. In addition, the trichomes are usually smaller.

A few of the special, big forms grow high in the Andes of Ecuador. These have the largest trichome wings of any *Tillandsia*, so the plants are the most fuzzy in appearance. The wings are elongated on one side and perpendicular to the leaf surface. This is why they are so visible and the leaf surface so coarse in appearance.**

The bloom scape is slender and usually exserts the lanceolate, complanate, compound digitate spikes beyond the leaves; however, there are exceptions. The glabrous, imbricate floral bracts are purple, which contrasts strikingly with the snowy white leaves. The small petals are aster violet, although the petals of some Peruvian populations have white tips.

Tillandsia tectorum often grows in areas subject not only to intense sunlight but to high levels of ultraviolet radiation as well (because of high elevation). The large trichomes reflect most of this, protecting the epidermis. This should be kept in mind when choosing a location for the plant. The grower must realize that the leaves will reflect much of the light impinging on them, so they must eventually be subjected to strong light. This can be accomplished by placing the specimen in a more exposed or higher position relative to other tillandsias.

Tillandsia tectorum lives in oligotropic (nutrient deficient) environments. Since there are so few nutrients available, it has evolved a very slow rate of growth. Other factors being equal, a plant that grows slowly does not need the nutrients of a plant growing in an eutropic (nutrient abundant) environment.

When *Tillandsia tectorum* is subjected to substantial and frequent moisture, much of the attractive trichome tomentum becomes transparently "glued" to the epidermis and disappears. Because of this, the leaves become more green and glabrous. Also, the blades may become stuck together as they grow out from the center of the plant. This species should be watered less frequently than others.

If placed in an interior location that does not receive high light levels, *Tillandsia tectorum* will often survive longer than other species. Of course, it will not grow at all in low light conditions. The key for long-term indoor growing is to give the plant bright light, and water it sparingly but thoroughly every one to three weeks.

*Smith, L.B. & Downs, R.J. 1977. *Flora Neotropica, Monograph, No. 14, Part 2, Tillandsioideae*. New York, N.Y. Hafner Press, pg. 786.

***See* subsection on Trichome Variations, pg. 208.

A.

B.

A. The largest form of *Tillandsia tectorum* from high in the Andes of Ecuador.

B. Specimen of *Tillandsia tectorum* with inflorescence exceeding the leaves.

Tillandsia tenuifolia

Linnaeus; 1753
(teh-noo-i-FO-lee-a)
subgenus: *Anoplophytum*

***Tenuifolius* is a combination of the adjective *tenuis,* which means "slender" or "thin," and the adjective *-folius,* which means "leaved."**

Tillandsia tenuifolia, one of the earliest described species, is widely distributed, ranging from the West Indies to Bolivia and Argentina. It grows epiphytically in forests, often in huge clumps. *Tillandsia tenuifolia* is variable and has at least five described forms. The species is found from near sea level to over 2,700 meters in elevation.

An average adult specimen of *Tillandsia tenuifolia* is about 8 cm. across and 12-15 cm. high. The species is caulescent and is at once attractive, durable, and inexpensive. The relatively short, polystichous leaves often give the appearance of pine needles and are usually 4-8 cm. long. They are green, which indicates a habitat of forest woodlands instead of the more exposed environment of scrubland or desert. For this reason, the plant prefers locations that are brightly lit but not subject to much direct sunlight.

The minutely appressed-lepidote blades are narrowly triangular and

D.

subulate-attenuate. Because of their thinness, the leaves will dehydrate more quickly than those of many other plants that are more drought tolerant and/or succulent.

The glabrous bloom, although not as large as similar species such as *Tillandsia aeranthos* and *T. stricta*, is striking nonetheless. It is simple and about equal to the leaves in length. The imbricate floral bracts of Brazilian and Argentine plants are ruby red, and the flared, polystichous flowers are white or plum purple depending on the variety. In California, *Tillandsia tenuifolia* blooms regularly between December and February for a month or so.

Like many tillandsias, once the plant has produced pups after blooming, the parent gradually fades and is overgrown by acrotonal offsets. If desired, in lieu of growing a specimen clump, the pups can be split into separate plants when they have grown to almost adult size. To do this, simply pretend the plants make a wishbone. If held by their bases, they can be gently separated. Each individual will have its own wire-like roots, and these will split the stem and come away with the respective plant. An alternative method is to snip them apart with scissors or a sharp knife at the point of connection.

Tillandsia tenuifolia may be mounted upside down, horizontally, or vertically, and it can be cultivated in many different situations outdoors. One should be sure that the plant receives water commensurate with the temperature, humidity, and level of light. Fertilization during the warm months promotes faster, larger growth. *Tillandsia tenuifolia* is an easy species to care for and is prolific enough after blooming to grow into a nice clump within three or four seasons.

A.

B.

C.

A. Blue-flowering form of *Tillandsia tenuifolia* from Brazil.

B. An interesting, brightly colored form of *Tillandsia tenuifolia* that was presented to the author by K. Yamaguchi in Aichiken, Japan, in 1977.

C. The white-flowering form of *Tillandsia tenuifolia* from Argentina.

D. *Tillandsia tenuifolia* thriving in a submesic habitat in Argentina.

Tillandsia tricolor

Schlechtendal & Chamisso; 1831
(TREE-cull-or)
subgenus: *Tillandsia*

***Tricolor* is a combination of the prefix *tri-*, which means "three," and the word *color*, which has the same meaning as in English. The epithet refers to the three colors of the inflorescence—cherry, canary, and indigo.**

Tillandsia tricolor usually grows as an epiphyte in woods in Mexico and Central America. It is a stemless species that averages about 30 cm. in diameter and 20-25 cm. in height. *Tillandsia tricolor* forms a dense rosette. The leaves are covered with a layer of widely dispersed, minutely appressed trichomes that give the plant a subglabrous appearance. The sheaths are casteneous, and the blades range from yellow-green to red. The blades are linear-triangular, long-attenuate and relatively stiff. With bright light and high humidity, *T. tricolor* will become a colorful standout.

The inflorescence can be simple or compound. The suberect to spreading spikes are linear-lanceolate and complanate. The scape and primary bracts are cherry. The spikes are erect to spreading, linear-lanceolate, and complanate. The glabrous, imbricate floral bracts are distichous and vary from canary yellow to lettuce green. The flowers are indigo. *Tillandsia tricolor* normally produces numerous offsets, occasionally as many as a dozen.

Tillandsia tricolor is often confused with *T. punctulata* because it is similar in appearance. But there are differences that make these two species easily distinguishable. The bases of *T. tricolor* are brown, and those of *T. punctulata* are usually a maroon so dark as to appear almost black. In addition, virtually every specimen of *T. tricolor* has a series of casteneous dots as the sheaths blend into the green of the blades. The sheaths of *T. punctulata* are wider, the blades brighter green, and the tips recurve. The leaves of *T. tricolor* are stiffer.

Both species produce stoloniferous offsets, but the runners of *Tillandsia tricolor* are tougher and thicker. The bases of many mounted specimens of *T. tricolor*, themselves originally offsets, have cataphyllic leaves covering the base of the plant.

Tillandsia tricolor prefers bright light but not direct sun or arid conditions. However, since the leaves are more chartaceous (hard—the texture of stiff writing paper or parchment) than those of *T. punctulata*, they can withstand stronger light and longer periods of dryness. As with all tillandsias, the key to plant vigor is a combination of good light, water, and a fertilization program that gives the plant enough nutrients to maximize growth potential.

A.

B.

Tillandsia utriculata

Linnaeus; 1753

(oo-tri-coo-LAH-ta)

subgenus: *Tillandsia*

***Utriculus* originally meant a "small leathern bottle." The suffix *-ata*, indicates possession or likeness. *Uticulata* is used botanically to mean "bladderlike" or "inflated." It was the earliest *Tillandsia* to be described; the capacity of a plant to impound water must have been novel.**

Tillandsia utriculata is found from sea level to nearly 1,200 meters. Its range extends from Florida to the Caribbean, Mexico, Guatemala, and Venezuela. This species grows epiphytically in trees, often in places that receive at least some direct sunlight.

Tillandsia utriculata forms a stemless rosette that grows up to a meter in diameter. Over a year's time, with good growing conditions, the plant will produce enough roots to anchor it if a proper mounting substrate is provided. With bright light, many of the leaves often have beautiful, deep maroon markings on them.

Tillandsia utriculata may be grown epiphytically or terrestially. As an air plant, it presents an imposing sight. Terrestially, *T. utriculata* looks more like other tank-

B.

type bromeliads. However, it will grow larger when potted because the roots will absorb the moisture and nutrients that are available in a good soil mix.

The leaf blades are long, linear-triangular, and attenuate. They are coriaceous but more flexible than those of species such as *Tillandsia fasciculata* and so are less subject to damage from shipping or handling. Mature leaves are spreading, the older ones recurving. Since the plant produces a large number of leaves, it is very attractive when well grown. The adaxial blade surface is subglabrous while the abaxial epidermis is finely covered with appressed trichomes.

The huge, compound inflorescence is often a lustrous carmine. The glabrous scape and rachis are tall, and the spikes are suberect to lateral. The flowers are pale chartreuse. The Floridian form of *Tillandsia utriculata* does not normally produce offsets; the Guatemalan plant usually does. There is also a viviparous form of *T. utriculata* that produces offsets on the inflorescence.

In an immature state, *Tillandsia utriculata* can sometimes be confused with a mature *T. polystachia*. The leaves of *T. utriculata* are generally longer, more spreading, and more numerous. They are also brighter green, and while the leaves may blush with a colorful hue in bright light, they never have the carmine cast and dull leaf color of *T. polystachia*.

Tillandsia utriculata is a species that will grow well under almost any conditions. Bright light, frequent watering during dry and/or hot months, and a consisent fertilization program produce the best results. With relatively poor light and infrequent watering, the plant will thrive but not grow as rapidly. The leaves will be dark green.

Tillandsia vicentina

Standley; 1923

(vih-sen-TEE-na)

subgenus: *Tillandsia*

***Vicentina* is the Latin form for the Spanish word, *Vicente*. The type location is the volcano of San Vincente in El Salvador.**

Tillandsia vicentina is a stemless species that grows epiphytically in forest habitats at elevations of 1,400-2,700 meters. It is native to southern Mexico, Guatemala, El Salvador, and Nicaragua.

Tillandsia vicentina is a graceful plant once it has become established in its growing environment. The many long, slender leaves, which are delicate silver-green, gradually spread to form an attractive rosette. The blades are finely covered with velvety, silver trichomes. The casteneous sheaths complement the blades.

An average mature specimen is large, growing to approximately 50 cm. across and 30 cm. high. In the non-blooming state, *Tillandsia vicentina* is often confused with *T. fasciculata* but is distinguished from it by the softness and flexibility of the leaves. Those of *T. fasciculata* are harder and stiffer in comparison.

B.

The inflorescence of *Tillandsia vicentina* is normally digitately compound. The tall, slender scape often exserts the complanate bloom spikes above the leaves. The scape is pastel green and covered with an indumentum of fine, appressed trichomes that give it a whitish cast. The distichous, carmine floral bracts are densely imbricate and frosted with a layer of appressed, cinereous trichomes. The flowers are grape violet. *Tillandsia vicentina* usually produces one to three offsets.

Tillandsia vicentina is more susceptible to rot than some of the other species. It is also more sensitive to poor water, which can be defined as water either too alkaline or water that has become stagnant (which often seems to be the case). One should water this plant thoroughly and flush it well.

This species is generally easier to establish outdoors because the plant prefers air movement. If given sufficient light, humidity, and water, *Tillandsia vicentina* can be grown indoors as well.

Tillandsia vicentina should be cared for with medium to bright light and commensurate amounts of water and fertilizer. This species, like most others, will adapt to conditions that are far from ideal.

A.

A. Two blooming specimens of *Tillandsia vicentina* on their original tree branch. They grew from 5 cm. seedlings to maturity in four years at Rainforest Flora, Inc. They were grown outside under 30% shade cloth.

Tillandsia xerographica

Rohweder; 1953

(zehr-oh-GRAH-fi-ca)

subgenus: *Tillandsia*

Xerographica is a combination of the Greek adjective *xeros*, which means "dry," and the noun *graphicos*, which means "painting." The epithet is a reference to the inflorescence that appears as if colored with pastel chalks.*

Tillandsia xerographica is an epiphytic species that ranges from southern Mexico to Guatemala and El Salvador. The most common environment is desert. At these locations, cool nights often produce morning dew that wets the plants. The rest of the day is dry—so dry, in fact, that the areas in Guatemala are also used as natural kilns to dry wood.** In addition, the plants must sometimes survive nine months of the year with no rainfall.

Tillandsia xerographica is one of the most popular species. The stemless plant is large, showy, and unusual. The sheaths are broadly ovate or obicular; the narrowly triangular, gray-white leaf blades are spreading, decurving, and sometimes contorted. An average adult plant is approximately 25 cm. in width and height. However, large specimens can grow to be twice that size.

Tillandsia xerographica is a slow growing species, often developing for several years before reaching maturity. The advantage to slow growth is that the bloom is long lasting; from bud to spent bloom may be many months.

The compound inflorescence of *Tillandsia xerographica* is tall and colorful. The scape bracts vary from salmon to clear pink. Against the whitish leaves their colors are striking. The erect to suberect spikes have glabrous, imbricate, buttercup floral bracts. From these bracts emerge mauve flowers. After blooming, the inflorescence retains bright color for weeks before gradually turning brown.

B.

The leaves of *Tillandsia xerographica* are densely covered with appressed, cinereous trichomes. They reflect excess light, insulate the leaf from desert heat, and protect the photosynthetic chlorophyll pigment from photobleaching (sunburn). The trichomes also help the leaves absorb and retain moisture. Their responsibilities are quite different from those of another xeric species with pronounced trichomes, *T. tectorum*.***

This plant is difficult to overwater if it is receiving bright light; underwatering is possible. Often this is not noticed until the plant receives a heavy watering and soon thereafter falls apart. The hapless owner then attributes his bad luck to "overwatering" when, in fact, the opposite occurred.

Common mistreatment occurs when a grower allows foliar meristematic tissue to remain wet. This happens most frequently when the grower allows water in the cup of the plant to become stagnant. As with many bromeliads, this situation may provide an environment for eventually lethal forms of fungi and bacteria.

The leaves are relatively succulent because of the length of time this species must endure droughty periods. Interestingly, the appearance of the plant itself can change tremendously. When *Tillandsia xerographica* is subjected to moisture—either in nature or in cultivation—the leaves have the capacity to absorb large quantities. When this happens, they straighten and become more succulent. When grown under dry conditions, the leaves desiccate and become more twisted. This phenomenon also occurs with *T. streptophylla*. The plant is healthy in either state as long as damaging extremes are avoided. Some *aficionados* are more appreciative of the twisted leaves while others favor the lusher, more succulent condition. In either form, this species qualifies as a collection centerpiece.

*Werner Rauh; a conversation with Rohweder. 1985 personal communication with Rauh.

**William W. Harris; 1984 personal communication.

****See* section of Foliar Trichomes, pg. 206.

Tillandsia xiphioides

Ker-Gawler; 1816
(ziph-ee-oh-EE-des)
subgenus: *Anoplophytum*

***Xiphioides* is a Greek compound word. *Xiphos* is the noun for "sword," and *-ides*, an adjectival suffix, indicates a "resemblance." The epithet, which means "swordlike," refers to the lanceolate shape of the inflorescence.**

Tillandsia xiphioides is indigenous to Bolivia, Paraguay, Brazil, Uruguay, and northern Argentina. It grows epiphytically and saxicolously in dry habitats at elevations of 700-2,700 meters. This is one of the more popular species because the blooms are large and fragrant.

An average mature plant is caulescent and about 15 cm. in width and height. A large specimen grows to as much as 22 cm. The lepidote blades are erect to spreading, narrowly triangular and attenuate-subulate. The major leaf colors are silver, brown, and green. The leaves are neither as numerous nor the same silver-

C.

green as *Tillandsia ixioides,* a species with which it is often confused because of name similarity.* *Tillandsia xiphioides* produces roots readily and affixes itself to the growing substrate.

The simple inflorescence is lanceolate and the colors are pale beige and moss green—not particularly eye-catching. The distichous floral bracts are densely imbricate and glabrous. The iris-shaped flowers are relatively large, creamy white, and exude a strong citrus fragrance, unlike that of any other *Tillandsia*. Although the three to eight flowers are relatively few in number, they do produce great numbers of seeds in large pods when fertilized.

The yellow seen at in the center of the flower at anthesis (when the flower is in full bloom) are pollen grains given up by the anthers. The anthers are either barely included inside or exserted from the throat of the corolla. The pollen is often sticky, a result of mixing with the nectar used to attract its probable moth pollinator.

Tillandsia xiphioides is easy to cultivate although it is sometimes susceptible to rot. It needs to dry completely between waterings. When provided with bright light, water, and fertilizer, the plant grows and roots rapidly. *Tillandsia xiphioides* is a handsome species, certainly worthy of a front and center location in one's collection.

**See* description of *Tillandsia ixioides*, pg.66.

A.

B.

B. A piece of natural tree branch covered with specimens of *Tillandsia xiphioides*.

C. *Tillandsia xiphioides* in xeric Argentine habitat.

Tillandsia xiphioides v. *tafiensis*

L.B. Smith; 1970

(tahf-ee-EN-sis)

subgenus: *Anoplophytum*

***Tafiensis* is the latinization of the type locality for this variety of *Tillandsia xiphioides*. The area of Tafi is in the state of Tucuman, Argentina. The adjectival suffix *-ensis* is used to indicate origin or place.**

Tillandsia xiphioides v. *tafiensis* grows as an epiphyte or saxicole at elevations of 2,000-3,300 meters. It is endemic to northwestern Argentina where its distribution is limited.

The morphology of *Tillandsia xiphioides* v. *tafiensis* is similar to *T. xiphioides* but is most easily distinguished from it because of the heavy leaf tomentum. These trichomes have large, coarse wings that make them easily visible to the naked eye.

This plant is encountered in the Andes growing singly and in small clumps on granite outcroppings. The rare, miniature *Tillandsia aizoides* Mez (eye-zo-EE-des) often grows in the same clumps with it. Various brightly colored lichens

D.

also thrive on the granite, an indication of consistently humid growing conditions.

Tillandsia xiphioides v. *tafiensis* is normally smaller than *T. xiphioides*, an average adult measuring approximately 10 cm. across and 12 cm. in height. The leaves of *T. xiphioides* v. *tafiensis* are more erect and not as spreading. When mounted with the plant base close to a substantial substrate, *T. xiphioides* v. *tafiensis* will reward its owner with rapid root attachment.

Although smaller, the inflorescence is similar to that of *Tillandsia xiphioides*—large, lanceolate, and moss green and beige. The blooms are almost indistinguishable, but the corolla of *T. xiphioides* v. *tafiensis* is slightly more conical in appearance than that of *T. xiphioides* v. *xiphioides*. The blooms of both are large, white, and iris-shaped. The flowers of *T. xiphioides* v. *tafiensis* are also aromatic, but the scent is slightly different from that of *T. xiphioides*.

As with other xeric species, *Tillandsia xiphioides* v. *tafiensis* prefers bright light. But because of the large, dense tomentum of trichomes, care must be taken to ensure adequate ventilation when the plant becomes wet. The plant may suffocate quickly if left wet in stuffy, humid air for any length of time. This period is difficult to ascertain because other factors such as light levels and temperature have a determination in plant health.

A.

B.

C.

C. Note the dense tomentum and offset of *Tillandsia xiphioides* v. *tafiensis*.

D. Saxicolous *Tillandsia xiphioides* v. *tafiensis* and *T. aizoides* in the Andean foothills of Argentina.

Vriesea espinosae

Gilmartin; 1968

(VREE-say-a ess-pi-NO-sigh)

subgenus: *Vriesea*

The genus *Vriesea* was erected in 1843 by John Lindley for a Dutch professor of botany, H. de Vries (1807-1862). *Vriesea* espinosae is named for R. Espinosa, who discovered this species in Ecuador.

Vriesea espinosae is a tillandsia-like species that grows epiphytically in dry shrubs and woods in Peru and coastal Ecuador at elevations of 200-1000 meters. It is often found growing together with *Tillandsia disticha*.

What determines that *Vriesea espinosae* is a *Vriesea* instead of a *Tillandsia* has nothing to do with a xeric environment. *Vriesea* flowers have minute petal flaps on the inside base of the petals, which separates them from tillandsias.* In nature, the plants grow in large groupings that literally encircle tree branches. These clumps are usually kept at least two meters above the ground by grazing goats who eat the lower growing plants.

Vriesea espinosae is stemless and grows into a porcupinish, hemispherical rosette. The more common Peruvian form is 12 cm. wide and 9 cm. high while the larger Ecuadorian plant is about 30 cm. across and 20 cm. high. *Vriesea espinosae* has an attractive configuration because the stiff, linear-triangular, attenuate leaf blades are symmetrically arranged. The densely lepidote foliage is covered with appressed, cinereous trichomes that give the plant a silver-gray hue.

The erect inflorescence is tall, slender, complanate, and simple with imbricate, glabrous floral bracts. The foliage, crimson bloom spike, and grape violet flowers provide a colorful, dramatic contrast for the grower.

One of the most desirable traits of this species is the stoloniferous way in which it vegetatively reproduces. After blooming, *Vriesea espinosae* develops several, long stolons that are covered with tiny cataphyllic leaves. A new plant gradually grows at the tip of each stolon. The dead, woody remnants of the ancestral plant and its stolons often remain connected for three or four generations.

As a xeric plant, *Vriesea espinosae* tolerates periods of strong light, heat, and drought in the native habitat. In cultivation, it grows robustly with similar conditions but will grow more rapidly with a generous watering and fertilization schedule in hot, dry environments.

**See* section on Taxonomic Division of Tillandsioideae Genera, pg. 173.

A.

A. Peruvian form of *Vriesea espinosae*; plant shown slightly larger than average.

Chapter 3
Additional Photographs

Fig. 3.1

Fig. 3.2

Fig. 3.3

Fig. 3.1 *Tillandsia compressa* Bertero.
Fig. 3.2 *Tillandsia aizoides* Mez.
Fig. 3.3 *Tillandsia bryoides* Grisebach.

Fig. 3.4

Fig. 3.5

Fig. 3.6

Fig. 3.4 *Tillandsia chiapensis* Gardner.

Fig. 3.5 *Tillandsia capitata* 'Marrón.'

Fig. 3.6 *Tillandsia deppeana* Steudel.

Fig. 3.7

Fig. 3.7 *Tillandsia flabellata* v. *flabellata* Baker.

Fig. 3.8 *Tillandsia didisticha* Baker.

Fig. 3.9 *Tillandsia ehlersiana* Rauh X *Tillandsia streptophylla* Scheideweiler.

Fig. 3.8

Fig. 3.9

Fig. 3.10

Fig. 3.11

Fig. 3.12

Fig. 3.10 *Tillandsia geminaflora* Brongniart.

Fig. 3.11 *Tillandsia funkiana* Baker.

Fig. 3.12 *Tillandsia hondurensis* Rauh.

Fig. 3.13

Fig. 3.14

Fig. 3.15

Fig. 3.13 *Tillandsia leonamiana* Pereira.

Fig. 3.14 *Tillandsia ignesiae* Mez (inflorescence enlarged approximately 3 times).

Fig. 3.15 *Tillandsia magnusiana* Wittmack X *Tillandsia compressa*.

Fig 3.16

Fig. 3.17

Fig. 3.18

Fig. 3.16 *Tillandsia lindenii* 'Pink Plume.'

Fig. 3.17 Bloom of *Tillandsia lindenii.*

Fig. 3.18 Bloom of *Tillandsia lindenii* 'Pink Plume.'

Fig. 3.19

Fig. 3.20

Fig. 3.21

Fig. 3.19 *Tillandsia lucida* Morren.

Fig. 3.20 *Tillandsia mallemontii* Glaziou.

Fig. 3.21 *Tillandsia mitlaensis* Weber et Ehlers.

Fig. 3.22

Fig. 3.23

Fig. 3.24

Fig. 3.22 *Tillandsia pruinosa* Swartz.
Fig. 3.23 *Tillandsia neglecta* Pereira.
Fig. 3.24 *Tillandsia roseoscapa* Matuda.

Fig. 3.25

Fig. 3.26

Fig. 3.27

Fig. 3.25 *Tillandsia ponderosa* L.B. Smith.

Fig. 3.26 *Tillandsia schatzlii* Rauh.

Fig. 3.27 *Tillandsia sucrei* Pereira.

Fig. 3.28

Fig. 3.29

Fig. 3.30

Fig. 3.28 *Tillandsia vernicosa* Baker.
Fig. 3.29 *Tillandsia* X *victoria* Foster.
Fig. 3.30 Undescribed *Tillandsia* from Guatemala.

Fig. 3.31

Fig. 3.31 *Vriesea rauhii* L.B. Smith.

Fig. 3.32 An amazing, elongated form of *Tillandsia araujei* Mez presented to the author in 1985 by Renate Ehlers in Stuttgart, West Germany.

Fig. 3.33 *Tillandsia usneoides* Linneaus.

Fig. 3.32

Fig. 3.33

Fig. 3.34

Fig. 3.35

Fig. 3.36

Fig. 3.34 An exercise in self-strangulation by a *Tillandsia duratii* (with no human help)!

Fig. 3.35 Hummingbird nesting on a specimen of *Tillandsia duratii*.

Fig. 3.36 The power of offsets is demonstrated by this *Tillandsia duratii,* which pushed its way through a leaf of the mother plant.

Fig. 3.37

Fig. 3.38

Fig. 3.39

Fig. 3.37 An unexplained lateral inflorescence from the base of *Tillandsia xerographica.*

Fig. 3.38 An interesting natural phenomenon—a doubleheaded *Tillandsia stricta.*

Fig. 3.39 A small grasshopperlike insect pauses to survey the surrounding countryside on the spike of *Tillandsia xiphioides.*

Fig. 3.40

Fig. 3.41

Fig. 3.42

Fig. 3.40 A dragonfly minutely examines an involute blade of *Tillandsia caput-medusae.*

Fig. 3.41 With a deft touch, an adventurous tarantula knows it has reached the end of the line on *Tillandsia ixioides.*

Fig. 3.42 One leaf getting trapped inside another is a fairly common occurrence in *Tillandsia bulbosa*. The surrounded leaf can usually be extracted with care.

Fig. 3.43

Fig. 3.43 *Androlepis skinneri* Brongniart.

Fig. 3.44 *Abromietiella brevifolia* Castellanos.

Fig. 3.45 *Aechmea orlandiana* 'Ensign'.

Fig. 3.44

Fig. 3.45

Fig. 3.46

Fig. 3.47

Fig. 3.48

Fig. 3.46 *Billbergia rosea* Beer.
Fig. 3.47 *Bilbergia* Muriel Waterman.
Fig. 3.48 *Guzmania sanguinea* Andre.

Fig. 3.49

Fig. 3.50

Fig. 3.51

Fig. 3.49 *Neoregelia* 'Candy Stripes'.*

Fig. 3.50 *Catopsis subulata* L.B. Smith.

Fig. 3.51 *Puya* species from Peru.

*First use of this name.

Fig. 3.52

Fig. 3.53

Fig. 3.54

Fig. 3.52 *Neoregelia* 'Lambert's Pride.'

Fig. 3.53 *Neoregelia* Doris.

Fig. 3.54 *Neoregelia compacta* L.B. Smith.

Chapter 4
Tillandsia Biographical History

Tillandsia Biographical History

This section will give the reader a sense of the history and tradition that have brought bromeliads in general, and tillandsias in particular, to their present state of interest. Unfortunately, space will allow only a brief glimpse at some of the adventurers, collectors, systematists, and authors who have built the rich collection of species and data now available. I have written about a few of the people who either have described

Fig. 4.1 Gaspard Bauhin

species in the book or have been instrumental in the popularization of bromeliads.

Bromeliads were first seen by Europeans on the initial voyage of Christopher Columbus (1451-1506) in 1492. Besides being one of history's great navigators,[1] Columbus was also a sensitive naturalist who was fascinated by all he saw in the New World.* On one visit ashore he remarked in his journal that he had encountered an extraordinary tree that had "branches of different kinds, all on one trunk, and one twig is of one kind and another of another, and so unlike that it is the greatest wonder in the world. . . .For instance, one branch has leaves like a cane, others like mastic; and thus on one tree five or six kinds, and all so different."[2]

Obviously, no tree approaches such polymorphy; Columbus must have seen a tree full of epiphytes, including tillandsias, which are ubiquitous on those Carribean islands.

Ananus comosus, the pineapple, was used as an important foodstuff by the Carib Indians on the island of Guadeloupe. The Indian name for the pineapple was *Karatas*. Bromeliads came to the attention of mainland Europeans after the second voyage of Columbus, in 1493. The plant found an enthusiastic reception, and by 1549, the pineapple was being commercially produced in other countries.

The taxonomic history of *Tillandsia* began in 1623 when Swiss physician and botanist, Gaspard Bauhin (1560-1624), published *Pinax Theatri Botanici.* It was the first work devoted to the organization of plants. There were over 6,000 entries, expertly categorized. One of them was a short description of a plant that, over a century later, was to be described by Linnaeus as *T. utriculata.*

In those early days, any air plant was referred to as a form of mistletoe, the only other "air plant" with which Europeans had had experience.

Charles Plumier

Priest, botanist, and early explorer of the West Indies, Frenchman Charles Plumier (1646-1704) first visited Hispaniola in 1689. Later he visited the West Indies on a second voyage as "Botanist to the King." As an avid scholar and artist, he sketched the plants he saw. He also compiled descriptions and was one of the first persons to use the binomial method of nomenclature. Many of his original drawings remain unpublished in the ar-

Fig. 4.2 Charles Plumier

chives of the Museum of National History in Paris.

Plumier described *Tillandsia paniculata*, an immense tank-type species, in his *Nova Plantarum Americanarum Genera* (1703). The description read, *"Renalmia ramosissima, floribus variegatis & circinnatis."* Linnaeus later accepted this description in his *Species Plantarum* (1753) and gave it the name of *Renealmia paniculata*. In his second edition, the description

*The term "New World" was coined by Peter Martyr, who, although he never visited them, was the first historian of the Americas.

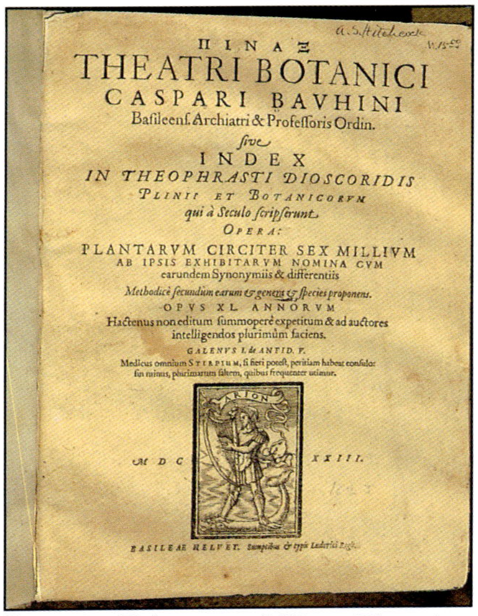

Fig. 4.3 Title page of *Pinax Theatri Botanici* (ΠΙΝΑΞ is the Greek equivalent for the Latin word PINAX and it means a "chart" or "register").

Viscũ baccis purpureis, in olea, Clus. hisp. & hist. qui & Visci oxycedri meminit.

III. Viscum racemosum.

Viscum polycoccon, Cam. ep.

Hoc à vulgari solùm baccarum racematim cohærentium copia differt.

IV. Viscum latifolium senis circulis utrinq; insculptis.

Viscum Indicum, Ad. Lob. Lugd. Tab. Ger.

V. Visci modo arboribus Indicis adnascens.

Peruviana alia Aloës facie, arboribus innascens Visci ritu, Ad. Lob. Lugd.

Planta Peruana Visco similis, Tab.

Viscum Peruvianum, Ger.

Fig. 4.4.

V. Growing like mistletoe on Indian trees.
Another, Peruvian, in the shape of an aloe, growing on trees like mistletoe, Ad.Lob.Lugd.
A Peruvian plant, similar to mistletoe, Tab.
Peruvian mistletoe, Ger.

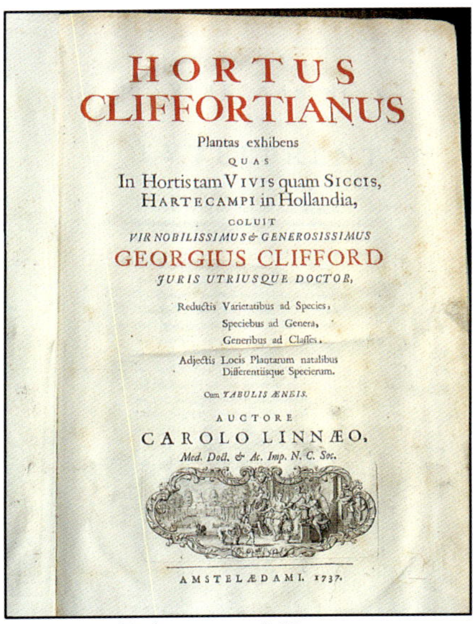

Fig. 4.5 Title page of *Hortus Cliffortianus.*

TILLANDSIA. *g. pl.* 283.

1. TILLANDSIA foliis basi conniventibus simul in utriculum.
Viscum caryophylloides maximum, flore tripetalo pallide luteo, semine filamentoso. *Sloan. flor.* 76. *hist.* 1. *p.* 188. *Raj. app.* 405.
Viscum indicum aliud. *Bauh. hist.* 1. *p.* 95.
Visci modo arboribus indicis adnascens. *Bauh. pin.* 423.
Peruviana alia facie, arboribus item innascens. *Dalech. hist.* 1829.
Crescit in truncis & ramis arborum frequens per Americam *calidiorem.*
Folia prope radicem arcte connivent & paulo supra basin coarctantur, adeoque utriculum constituunt in quem pluvia foliis patentibus incidens defluat & colligatur; collecta aqua servatur nec facile evolat evaporatque, cum utriculus superne angustatus sit. Hæc aqua præterquam quod plantæ conveniat & avibus, insectis & hominibus peregrinantibus urgente siti, nec præsente alia aqua, potum subministrat optimum in istis, ubi crescit, oris.
Caraguata nomen Plumieri americanum & barbarum est, in cujus locum substitui memoriam ELIÆ TILLANDSII, *primi, & unici Botanici, qui quondam in Finlandia floruit.*

Fig. 4.6

1. *TILLANDSIA,* its leaves closing together at the base into a utricule.
The largest mistletoe, caryophyloides, with a three-petaled pale yellow flower, seed covered with fine threads. *Sloan. flor. 76. hist.* 1.p. 188. Raj. app. 405.
Another, Indian mistletoe. *Bauh. Hist. 1. P. 95.*
Another, Peruvian in shape, likewise growing on trees. *Bauh. pin.* 423.
Like mistletoe growing on Indian trees. *Dalech. hist.* 1829.
It grows thick on trunks & branches throughout warmer America.
The leaves near the root close tightly, & a little above the vase they contract, & thus form a utricle into which rain water, falling on the open leaves, flows down & is collected; the collected water is kept & does not easily escape or evaporate, since the utricle is narrow at the top. In regions where the plants grow, this water, besides being useful for the plants, furnishes drink for birds & insects, & it is a good water supply for men traveling through when thirst is pressing & no other water is available.
The name, Caraguata Plumerieri, is American & barbarian. In its place (Tillandsia) is substituted in memory of ELIAS TILLANDS, *a great & unique botanist from Finland.*

Translation from the Latin courtesy of Father Charles Cavanach, S.J. Loyola Marymount University. Los Angeles, California. Commas have been added and the translation slightly updated in order to make reading easier.

was enlarged and the name *T. paniculata* was used for the first time.

Carolus Linnaeus

Carolus Linnaeus (1707-1778) wrote many botanical works of great importance. Among them were, *Genera Plantarum* (1737), *Hortus Cliffortianus* (1738), *Philosophia Botanica* (1751), and *Species Plantarum* (1753). Linnaeus made his principles and methods immediately usable all over Europe by writing in Latin instead of his native Swedish. He created a simplified form of Latin as the international language for the formal naming and describing of plants. He associated his works with an acceptable terminology and applied them in encyclopadic works, which became essential consulting tools for botanists, because no one else had assembled so much professional work in such a convenient mode of presentation.

Linnaeus built upon the botanical works accumulated over previous centuries. However, he did so much to organize the format and introduce terminology still in use that the differences between the botanical descriptions of Linnaeus and his immediate predecessors is actually greater than those of Linnaeus and modern botanists.

In 1737 a Dutchman, George Clifford, commissioned Linnaeus to describe the plants in his impressive botanical collection outside Amsterdam. The resulting work was titled *Hortus Cliffortianus*. It described the first *Tillandsia, T. utriculata*, which subsequently became the type for the genus (Fig. 4.6) .

Fig. 4.7 Carolus Linnaeus

Fig. 4.8 Alexander von Humboldt

In his 1753 publication of his *Genera et Species Plantarum*, Linnaeus wrote of 14 bromeliad species in the three genera of *Tillandsia, Bromelia*, and *Renealmia*, which he had constructed in *Hortus Cliffortianus*. All of these plants were new because they appeared at the beginning of the binomial method for naming plants. Curiously, Linnaeus did not make a single original description from specimens. Instead, he substituted binomials for the polynomials of previous botanists, especially the Frenchman, Plumier. He never made a trip to the Americas.

As the extent of the bromeliad family became known, the French explorer, botanist, and entomologist, Auguste Juame de Saint-Hilaire (1779-1853), elevated the plants to family status in 1805. He took the genus, *Bromelia*, and added the Latin suffix for families, *-aceae*.

Alexander von Humboldt,
Aime Bonpland,
Carl Sigismund Kunth

Explorer and scientist Alexander von Humboldt (1769-1859) gained worldwide renown in his time as the chief promulgator for the earth science studies of geography, geology, geomagnetism, climatology, and of particular interest to us, the relation of plants and animals to their physical surroundings, which is now known as the science of ecology.

"Alexander von Humboldt, indefatigable traveler, author, and father figure of science, was perhaps the most widely admired man of the 19th century. More places around the world are named for him than any other scientist."[3]

At the turn of the 19th century in the surroundings of Berlin, von Humboldt began to collect and learn to classify plant specimens. But the poor flora of the province of Brandenburg did not provide much stimulus for an ardent botanist, and Humboldt soon dreamed of journeys to more exotic lands.

In 1799 Humboldt's assistant, Aime Bonpland (1773-1858), accompanied Humboldt on an expedition to Central and South America. According to a number of authorities, the trip has long been accepted as the most important ever made to the Americas, because the results were so wide-ranging and long lasting in the fields of natural science.

The results of the immense quantity of specimens and observations related to the fields of botany, zoology, geology, geography, and history. They were amassed with almost incredible industry and often under great dif-

Fig. 4.9 Aime Bonpland

ficulty. Their collections and data served as the basis for study by many specialists. Humboldt and Bonpland ventured into areas that had not been previously scientifically investigated. They had the best equipment and used it expertly. Upon their return they wisely called upon the services of other highly qualified individuals to elaborate upon their findings.

Carl Sigismund Kunth (1788-1850) spent 22 years writing the comprehensive and systematic descriptions of the plants collected on this trip. Humboldt supplied notes on geographical distribution and supervised the cost and publication of the work. Bonpland supplied the specimens and notes, and Kunth, the author, had the scientific responsiblity for the work as published.

The original descriptions of *Tillandsia disticha, T. incarnata, T. secunda,* and *T. straminea* were published in their famous work, *Nova Genera et Species Plantarum.*

Diedrich von Schlechtendal
Adelbert von Chamisso

Diedrich Franz Leonhard von Schlechtendal (1794-1866) was an excellent systematic botanist and the author of numerous articles. While at Berlin he published a *Flora Berlinensis* in 1823-1824. Afterwards, Schlechtendal taught at the University of Halle and became director of the Botanic Garden.

He was editor of the renowned

Fig. 4.10 Carl Sigismund Kunth

Fig. 4.11 Diedrich von Schlechtendal

Linnaea from its first publication in 1826 until his death and was co-editor of the *Botanische Zeitung* (Botanical Newspaper). Few persons in Germany had a more extensive knowledge of plants or contributed more to the progress of botanical science during that period. Schlechtendal also constantly applied the resources of the gardens under his charge to close observations in systematic botany, the results of which were, from time to time, published in his journals.

Adelbert von Chamisso (1781-1838) was nine when his family escaped the terrors of the French Revolution by taking refuge in Berlin. A celebrated man of learning, he was one of the most gifted of the Berlin Romanticists. Some of his poetry was set to music by Robert Schumann. In addition, Chamisso wrote the Faust-like tale, *Peter Schlemihl's Remarkable Story* (1814), which won him lasting fame.

Chamisso first became acquainted with Schlechtendal in 1813 while studing botany and natural science. The two of them, along with others, went on long "herborizations," collecting and studying plants. He often wore "an antique garb, once the state dress of a South Seas Chief, much worn, mended and stained, with a black cap of cloth or velvet, a large green box suspended by leathern straps over his back, and a short pipe in his mouth, together with a rude tobacco pouch. And it may be supposed that when evening came and our Botanist was weary, travel-soiled, he did not make a very splendid appearance while bearing a pocket handkerchief crammed with plants. He met, on returning to Berlin, the *beau monde* of that capital all in their Sunday attire."[4]

In 1815 Chamisso went as naturalist on a three-year sailing expedition around the world, the main object of which was to discover the Northwest Passage. His diary, published in 1821, *Reise um die Welt mit der Romanzoffischen Entdeckungs Expedition (Voyage Around the World with the Romanzov Discovery Expedition*), became a classic of its kind.

As the voyage was completed he became convinced ". . . that many difficulties and disappointments must ever attend such expeditions, and that the limited accommodation of a vessel, with the peculiar ideas of the captain, and the chance of accidents and many other obstacles, are apt to defeat a naturalist's best intentions and endeavors."[5] All of this notwithstanding, Chamisso's energy and zeal enabled him to overcome many of these disadvantages and amass a rich collection.

On the expedition, Chamisso spent the month of October 1816 in San Francisco. Over 30 new species were collected and later published with Schlechtendal in *Linnaea*. The locality sited was

Fig. 4.12 Adelbert von Chamisso

"ad portum San Francisco" and all were collected in the hills and downs about the Presidio. Very few, if any, are to be found there today.[6]

After receiving his doctorate of philosophy degree in 1819, Chamisso went to work at the Royal Herbarium in Berlin, where he worked for many years with Schlechtendal. In 1831 they published the descriptions included in this book for *Tillandsia punctulata, T. tricolor,* and *T. filifolia*.

Johann Georg Beer

An Austrian, Johann Georg Beer (1803-1873) spent some of his adult life as a councilman for the city of Vienna, and in the beginning was active in botany only as an amateur. His chief interest was in finding ways to better grow bromeliads and orchids. His collection was extensive because of his friendships with other botanists, and, after his interest took a more professional turn, he eventually became director of the Berlin Botanical Garden.

His first publication was *Praktische Studien an der Familie der Orchideen (Practical Studies of the Orchid Family)*, in 1854. This was followed in 1857 by *Die Familie der Bromeliaceen*, which was printed in Vienna. In 1863 he published *Beitrage zur Morphologie und Biologie der Familie der Orchideen (Contributions to the Morphology and Biology of the Orchid Family)*.

Beer's *Die Familie der Bromeliaceen* was the first monograph devoted to bromeliads. He attempted to arrange all the bromeliads then known into a taxonomic system based on phenotypic differences among species. He collected antecedent records and "added numerous new species described from living material. . . . He recognized 31 genera and 237 species, most of which are good today."[7] He did not base the work upon differentiation of floral characteristics—the method adopted by later botanists.

The following quote from the forward of his monograph will give the reader a peek at his *modus operandi* and the great challenge he faced with such dedication:

"Bromeliads are one of the most interesting plant families and include a great variety of forms. And how difficult was their classification at first. Although I often devoted myself completely to my multiple studies and was lost in a labyrinth of various concepts, I soon realized that industriousness and patience would lead to an admirable goal. In order to avoid having a great number of books in front of me all the time, there was only one thing for me to do—carefully copy all the illustrations of bromeliads available in the botanical literature. In doing this I achieved a truly useful work; I also studied all the living plants from this family available to me. They were the basis of the classification and the descriptions. Anyone who knows the confusion caused by the drawings will excuse me for daring to rely on my direct observation in creating the divisions of this plant family and everything down to the description of the final species. . . .We took time to acquire a great number of works yielding a rich treasure of *Bromeliaceae*. But here again we found the most regrettable misunderstandings:— I recall, for instance, Arrab's *Flora Fluminensis*! Other expensive works such as *La Bonite, Atlas Botanique* by Gaudichaud, and *La Coquille* by Duperrez lack, as does the first named work, any descriptions! In my method of classification, in which the total growth form is considered, such works, even though they have no descriptions nor good diagnoses, are nevertheless quite useful because there is attention given to the inner structure of the flower. . . .It is my hope that this work will prove that there are various roads leading to an understanding of plants; perhaps even new, useful, practical views will be opened up. I am quite convinced that many monocot plant families can be studied clearly with my method, keeping the characteristics, but in every case the overall shape of the plant must be the basis of the study."*

Fig. 4.13 August Grisebach

August Grisebach

In his youth, August Heinrich Rudolf Grisebach (1814-1879) was encouraged in the study of natural history by his uncle, botanical professor Georg Friedrich Wilhelm Meyer (1782-1856).

Grisebach entered the university of Gottingen in 1832 and studied botany as well as medicine. He then moved to the University of Berlin where he became a doctor of medicine in 1836. He spent his later career at the herbarium in Goet, Gottingen, where he became director of botanical garden.

Alexander von Humboldt's travels inspired him, as they did Charles Darwin. In March 1839, Grisebach set out on the most important journey of his life, a botanical expedition to the Balkan Peninsula. "The scientific results of this journey across Thrace, Macedonia, and Albania, made when he was 25 years old, entitle it to rank in botanical history with the celebrated journey into floristically unknown Lapland made a little over a hundred years earlier by Linnaeus at the age of 25."[8]

Over the years Grisebach became the world's leading authority on plant geography, partly through his own field work, partly by the study of herbarium collections from many lands, and partly through analysis of innumerable

*The author wishes to acknowledge and thank the late Wilhelm Weber, of the German Democratic Republic, who supplied much of the background material, and Harvey Kendall of Los Angeles for translating the German.

floristic and geographical publications. Grisebach's *Flora of the British West Indian Islands* (1859-1864) was the first of a series of British colonial floras initiated at Kew by Sir William Jackson Hooker (1785-1865).

He received and described the large plant collections of Argentines Paul Gunther Lorentz (1835-1881), for whom he named *Tillandsia lorentziana*, and Georg Hieronymus (1846-1921). He catalogued their plants in *Plantae Lorentzianae* (1874) and *Symbolae ad Floram Argentinam* (1879).

Fig. 4.14 Edouard Morren

From 1841 to 1856 Grisebach published yearly reports on plant geography and systematic botany. In 1872 he brought together the fruits of this study in his monumental, two-volume *Die Vegetation der Erde nach ihrer klimatischen Anordnung (The Vegetation of the Earth According to Its Climatic Zones)*. This and *Flora of the British West Indian Islands* are his most important single contributions to botanical literature.

Edouard Morren

Edouard Morren (1833-1886), son of the distinguished Belgian professor of botany and bromeliad enthusiast, Charles Morren, became interested in plants at an early age and decided to follow in his father's footsteps.

On his way to that goal, he received a well-rounded education, majoring in philosophy and the fine arts. He received his doctorate in the natural sciences and continued his horticultural studies by visiting many of the famous European botanical gardens.

Morren became interested in plant illustration as a teenager. His goal was to reproduce plants as realistically as possible. Great detail was paid to the fine nuances of light and shade. Hundreds of his magnificent illustrations appeared in *La Belgique Horticole*, which had been started by his father (*see* frontspiece to this chapter).

Besides being a professor of botany, Morren was director of the botanic gardens at Leige and worked closely with botanists around the world. Over the years he established a tremendous collection of bromeliads, both at his home and at the botanical gardens. He described 77 bromeliad species new to science. He also described plants in other families, most notably the Marantaceae and Orchidaceae.

Two of his prize pupils were Edouard Andre and Carl Mez. Of Morren, Andre said, "He had willingly instilled in me his own passion for the Bromeliaceae."* Mez said that he owed more to Morren than any other bromeliad botanist. Much of the interest in bromeliads today is a direct result of Morren's unflagging devotion to bromeliads—manifested in his writings, paintings, and teachings.

Morren enjoyed life, loved people, and was an attentive, generous man. His son, E. Morren, became a noted hybridist in his own right. Unfortunately, Edouard Morren drove himself hard and died at the relatively young age of 53. He left an uncompleted monograph of the *Bromeliaceae*. In addition, he was never able to realize his lifelong ambition of collecting in the jungles and forests of Brazil.

John Gilbert Baker

John Gilbert Baker (1834-1920) developed an early interest in plants, but he left school at 14 to help in his father's drapery and grocery business, which he worked at for the next 18 years. However, he continued to collect plants. At 15 he contributed an article to the periodical Phytologist, and his publications continued unabated for the rest of his life. His studies were broad, as evidenced by a review of the British roses published in the Naturalist in 1864 and a monograph of British mints published in the Journal of Botany in 1865.

Fig. 4.15 John Gilbert Baker

In 1864, a disastrous fire destroyed not only Baker's botanical library but his herbarium as well. Friends suggested that he leave his business practice and devote himself exclusively to botanical pursuits. In 1862 a large herbarium had been willed to Kew. The incorporation of this work on vascular crytograms was a matter in which the director of Kew, Sir William Hooker, was directly involved. The excellence of Baker's work with the roses, along with his current studies of ferns, prompted Hooker to invite Baker to assist in the arrangement of the material.

In 1866 the post of first assistant in the herbarium was created specifically for Baker, allowing him to permanently join the Kew staff. In addition to his work at the herbarium, he supplemented his income by teaching. In 1869 he was appointed lecturer on botany at the London Hospital Medical School, a post he held until 1881.

*Andre E.F. 1983. **Bromeliaceae Andreanae**. Berkeley, California. Big Bridge Press, Pacifica: Twowindows Press. pg. xxiv.

In 1882 he was appointed by the Society of Apothecaries to their lectureship on botany at the Chelsea Physic Garden.

In June of 1890 Baker was promoted to the keepership of the library and herbarium, and served in this post until his own retirement at age 65 in 1899. He continued to teach, "greatly to the advantage of his pupils"[9] for five years following his retirement.

Baker became a leading authority of his day on monocotyledons. His monograph, *Handbook of the Bromeliaceae*, was the first written in English and appeared in 1889. Baker's source material came from the living collections at Kew, the studies and paintings of Edouard Morren of Belgium, and the collections of A. Glaziou from the environs of Rio de Janeiro. Much of his work also was done with herbarium specimens. His monograph on bromeliads included 800 species that were grouped into 31 genera.

Baker's work as a naturalist and systematist brought him much recognition throughout his life, among them the Victoria Medal of Honour in Horticulture in 1897 and the Linnean Medal in 1899. He was elected to honorary membership of many national and international horticultural societies.

Baker's sense of proportion, which made him such a distinguished systematic writer, also enabled him to be an equally effective teacher. "He used the natural history of plants as a means of educating those on whom he inculcated the importance of botanical knowledge in everyday affairs. His style was lucid and concise, while he possessed the happy gift of ability to emphasize the salient features of his subject without neglecting its details. Baker's published works ensure the perpetuation of his memory as the last of a singularly gifted circle of systematic botanists. While any of them survive, those who worked with or were taught by Baker will cherish the recollection of one of the kindest and best of men."[10]

Edouard Francois Andre

Edouard Francois Andre (1840-1911) developed a keen interest in plants at an early age. His father was a horticulturist. At the age of 20 he was given the important job of head gardener for the city of Paris, a post he held for eight years. In taking an active part in

Fig. 4.16 Edouard Francois Andre

the creation of many of the city's fine public parks and gardens, Andre showed his skill as a landscape designer.

He next entered, and won, a competition to design a 370-acre park for the city of Liverpool, a project that lasted five years. This accomplishment brought him recognition, and he spent the next few years designing some of the most famous parks in Russia, Austria, Holland, England, Denmark, Italy, Switzerland, and Bulgaria. Andre created the gardens of Funchal in Madeira, Spain, the public gardens in Monte Carlo, the gardens of the Villa Borghese in Italy, and the terraced gardens and promenades in Luxembourg.

In 1860, while immersed in these other projects, he became a contributor to *Revue Horticole*. From 1870 to 1880 he served as editor of *L'Illustration Horticole*. In these journals he described a large number of bromeliads and other plants that have since become quite popular. Between 1864 and 1871 he wrote a number of well-received books.

In 1875 Andre was commissioned by the French government to lead an expedition to Colombia and Ecuador to study the vegetation in that region. He wrote a fascinating account of the journey that was published in 1889.* A 1983 deluxe English reprint is fast becoming a collector's item.** Andre's trip roughly followed the path of the Humboldt and Bonpland expedition of 1801-1802, and it covered every type of terrain. Of the many plants Andre collected, one was a small *Tillandsia* with a solitary red bloom. Edouard Morren later gave it the name of *Tillandsia andreana*.

After his return to Europe, Andre devoted much of his time to bromeliads and writing. He continued to write for *Revue Horticole* and *L'Illustration Horticole* and made one last trip to Latin America, to Uruguay, in 1890.

"Andre was greatly esteemed not only by his countrymen but also by citizens of every nation with whom he had contact. Perhaps the greatest distinction which fell to him was his appointment to the Professorship of Garden Architecture at L'Ecole Nationale d'Horticulture of Versailles in 1892. The chair—the first of its kind— was established on his behalf. This was just one of many honors bestowed upon him. He was made a corresponding member of the Royal Horticultural Society of London, an Officer of the Legion d'Honneur, and Officer of the Order of Leopold of Belgium, and he held many other distinguished orders."[11]

Carl Mez

Carl Mez (1866-1944) was, along with Edouard Andre, a prize pupil of the great Belgian botanist and taxonomist, Edouard Morren. Many of his later ideas came directly from Morren. After studying the natural sciences, especially botany, from 1883 to 1887, he moved and continued his studies in Berlin. He graduated with a doctorate in philosophy in 1888 at the age of 22. In 1899 he became professor for Systematic Botany and Pharmacognosy at the Univer-

*Andre, E.F. 1889. *Bromeliaceae Andreanae*, Paris, France. Librairie Agricole.

**Andre, E.F. 1983. *Bromeliaceae Andreanae*. Berkeley, California. Big Bridge Press, Pacifica: Twowindows Press.

sity of Halle. He worked there until 1910. He then left to become director of the Botanical Garden of Konigsberg, East Prussia, which is now Kaliningrad, U.S.S.R.

His published works were numerous. Those relating to bromeliads were: *Bromeliaceae in Maritus Flora Brasiliensis*, 1894;

Fig. 4.17 Karl Mez

Bromeliaceae in DeCandolle Monographiae Phanerogamarum, 1896; *Physiologische Bromeliaceen Studien*, 1904; *Monographia Bromeliaceae in Englers Pflanzenreich*, 1935.

Unlike many of his contemporaries, Mez took advantage of the opportunity to visit other botanical gardens and herbaria. He studied the work of Baker and Andre firsthand and borrowed material from far and wide.

According to Lyman Smith, Mez's first monograph on Bromeliaceae, *Das Pflanzenreich* (1896), was great, not so much in the number of species, which totaled 997, but in the clarity and thoroughness of its organization. It was the first treatment of the family that is still useful for taxonomic purposes today. His second tome, *Pflanzenreich* (1935), included 1,516 species.

In the introduction to his *Flora Neotropica* (1977), Smith says of Mez, "He brought order out of near taxonomic chaos and his system was basically so sound that the changes embodied here are largely quantitative. . . .There have been numerous minor changes, but his major concepts have stood the test of time".

In 1904 Mez published a long article summarizing a series of brilliant experiments and demonstrations designed to discover how bromeliad trichomes absorb salts and moisture with their leaves. "His reports were so thorough and his suppositions so convincing," Benzing wrote, "that little elaboration and refinement of this classic work was attempted until the 1970's, even though much more powerful analytical tools had become available some years before."[12]

Mulford B. Foster

Mulford B. Foster (1888-1978) has done as much or more than any other individual to popularize bromeliads in America. His unstinting devotion and enthusiasm led to the discovery of many new ornamental species on his frequent journeys to Latin America. His prolific pen spread the "bromel gospel" far and wide.

As an adult, Foster represented a tree company for five years in the states of Virginia, Maryland, and Washington, D.C. Subsequently, his naturalist and artistic nature led him to photography and painting at the age of 36. He moved to Orlando, Florida, in 1923 and became a landscape architect. Mulford married the former Eva Racine Sarasy in 1935.

At that time the Fosters began their plant-collecting excursions throughout Latin America. During these years, Mulford Foster collected over 200 new bromeliad species that were unknown to botanical science—an astounding feat. He has many plants named in his honor, including the bromeliad genus, *Fosterella*.

"A collecting trip with Mulford Foster to Jamaica was a revelation to me insofar as his knowledge of plants was concerned. He could identify every plant that he saw, no matter how mean or unobtrusive, and from a distance of at least 100 yards, he could distinguish a *Pitcairnia* from a grass. When he saw a bromeliad on a tree, he did not wait for the boys to collect it, but was the first one up, chuckling in glee that here was another treasure."[13]

Mulford Foster was one of the cofounders of the Bromeliad Society, Inc., in 1950. He was president of the Society from 1950

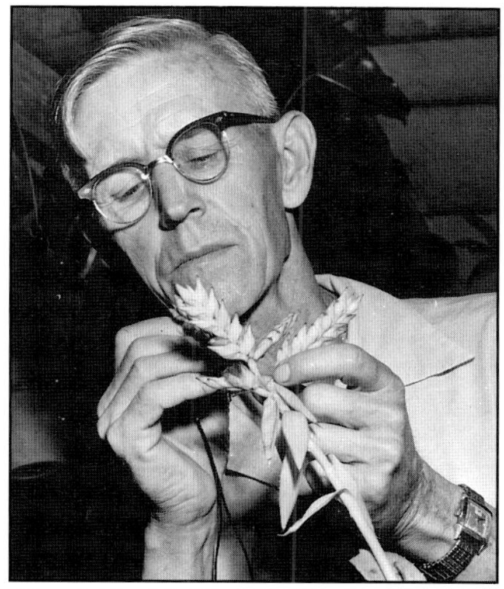

Fig. 4.18 Mulford B. Foster

to 1959, and editor of its journal from 1951 to 1958. During this time he contributed over 180 articles. Racine and Mulford's book, *Brazil, Orchid of the Tropics* (1945) is the fascinating account of "the agonies and the ecstasies" of traveling and collecting in Brazil in those days. In addition, Foster authored *Puya, the Pineapple's Andean Ancestor* for *National Geographic* magazine in October 1950; also, much of the first *Bromeliad Society Cultural Handbook* was his work.

Alberto Castellanos

Alberto Castellanos (1896-1968) was a botanist, collector, writer, horticulturist, and the foremost authority of his day on the bromeliads of Argentina. In addition, he was an honorary trustee of the Bromeliad Society.

Castellanos helped produce the most beautiful work on bromeliads to date. Bromeliads were included in Volume III of the seven-volume work on the flora of Argentina, commissioned by Juan Peron. *Genera et Species Plantarum* was published in 1945 in a limited folio-size edition of 1,500 copies that measure 33 cm. by 51 cm. and weighs nearly nine kilos. The 14 genera of bromeliads containing 99 species are illustrated by 101 plates. Fifty-seven are presented in full color and the accuracy and renditon of the plants is unsurpassed. In the text, each species

Fig. 4.19 Alberto Castellanos

has the original description followed by Castellanos, own more detailed explanation in Spanish. "Thanks to this great work the Bromeliaceae of Argentina are better known.and easier to identify than those of any other country in the world."[14] Very rare, the work is highly sought by bromeliad collectors.

In 1958 Castellanos left Argentina for a teaching position in the Museu Nacional in Rio de Janeiro. Although he retired in 1966 at the age of 70, he continued his work unabated. He helped to organize and build the Herbarium Bradeanum in Rio. At the same time, he continued to travel and collect. He was stricken ill returning from a field trip and died on September 5, 1968. After his death, a grove of trees was planted on the grounds of the Museu Nacional in his honor. Internationally renowned botanists planted the first trees. Lyman B. Smith represented the United States.

Lyman B. Smith

Lyman B. Smith (1904-) has always felt that had he not become a taxonomist, he would have been a language professor. He had the equivalent of seven years of French in high school and college.

After having received his degree from Harvard in 1925, Smith spent a year working for his father, but his real interest was taxonomy. He commenced his graduate studies in 1926. His thesis subject was the *Tillandsia* subgenus, *Pseudo-Catopsis* Baker. After his first two years of graduate school, he received a fellowship that took him to eastern Brazil in 1929. There, for the first time, he had the opportunity to observe and collect bromeliads in their natural habitats. Brazil became the target of his floristic interest and he returned there many times. He also made brief trips to Cuba and Costa Rica and several to Argentina.

From 1931 to 1947 Smith was a staff member at the Gray Herbarium of Harvard University. Prior to 1935, he made three trips to Europe where he did considerable work at the British Museum and Kew Gardens, at the botanical museums of Brussels and Liege in Belgium, and in Paris, France. While abroad, he examined the older herbaria collections and photographed the original types of many of the first bromeliads collected. In 1947 he joined the staff of the Botanical Department of the Smithsonian Institution, where he is currently Curator Emeritus of Botany.

Lyman Smith's publications include more than 300 titles in professional serials and popular journals. Some of the more prominent

Fig. 4.20 Plate LXXVII from *Genera et Species Plantarum*. Note detailed drawings of floral appendages.

are: *Bromeliaceae, North American Flora* (1938); *The Begoniaceae of Colombia* (1946); *Bromeliad Malaria* (1953); *Revisio Violacearum Columbiae* (1954); *Studies in the Bromeliaceae I-XVII* (1930-1954); *The Bromeliaceae of Brazil* (1955); *A Synopsis of the American Velloziaceae* (1962);

Fig. 4.21 Lyman B. Smith

Origins of the Flora of Southern Brazil (1962); *Solanaceas, Flora Ilustrada Catarinense* (1966); *Xyriadaceae, Flora of Brasilica* (1968); *Bromeliaceae, Flora de Venezuela* (1971); *Notes on Bromeliaceae* I-XXXIV (1953—1974). In addition to all of this, he has spent more than 50 years compiling and publishing information on the *Bromeliaceae*. His three-volume work, *Flora Neotropica* should stand for decades as the central reference point for the family.

Werner Rauh

Werner Rauh (1913-), along with Lyman Smith, has long been considered the world's leading authority on the *Bromeliaceae*. He has traveled and collected extensively in the Americas, and has written a number of books on the subject. He has also contributed regularly to the *Journal of the Bromeliad Society* for nearly 20 years.

Rauh grew up with a keen interest in flora and fauna, geography, geology, and the early history of his homeland. He studied in Leipzig, Innsbruck, and Halle, where he received his doctorate. He began publishing at the age of 24, and at 27 he moved to the University of Heidelberg.

Although the war took away important years of teaching and research, he began lecturing again at the university soon after its restoration. At this time he also went on expeditions across the Alps to the Mediterranean and North Africa. In 1952 he published his *Vegetation in the High Atlas Mountains*, a result of these trips.

In 1954 Rauh was able to realize an old dream, to explore the Andean vegetation of Peru and Ecuador; a second journey followed in 1956. On these trips he developed an interest in bromeliads and cacti.

In 1958 Rauh became the director of the Institute of Systematic Botany at the University of Heidelberg, a position he held until his retirement in 1981. From 1960 to 1982 Rauh, who was sometimes accompanied by his wife Hilda, went on numerous collecting trips to countries such as Madagascar, Africa, Saudi Arabia, Israel, New Guinea, and made many trips to the Americas. The result is that the Botanical Garden of the University of Heidelberg has become world renowned. The scientific significance of these trips is staggering. His discoveries, scientific descriptions, and publications are voluminous.

Rauh has received many accolades. He is an honorary member of the Bromeliad Society. He was elected president of the International Organization for the Study of Succulent Plants. He is an honorary member of almost all succulent societies around the world. In 1978 he received the *Cactus d'Or* (Golden Cactus) from her highness, Princess Grace of Monaco. In London he was honored in 1974 with the *Golden Veitch Memorial Medal* of the Royal Horticultural Society. In 1979 he received the *Willdenow Medal* from the Berlin Botanical Garden and Museum.

As of 1987, Rauh is still active, participating in collecting expeditions, attending international plant symposiums, and describing new species.

Literature Cited:

1. Morrison, S.E. 1977. *Admiral of the Ocean Sea.* Franklin Library. Franklin Center, Pennsylvania, pgs. 194-196.

2. *ibid.* pg. 242.

Fig. 4.22 Werner Rauh

3. McIntyre, L. 1985. Humboldt's Way. National Geographic Society Magazine. Vol. 168, No. 3. Washington, D.C., pg. 318.

4. Schlechtendal. 1866. Memoir of Chamisso. The London Journal of Botany. Vol. 2, pg. xxvii. Translated from the original *"Linnaea, Ein Journal fur die Botanik."* 1843.

5. *ibid.*

6. Eastwood, A. The Botanical Collections of Chamisso and Eschscholtz in California. Leaflets of Western Botany; Vol.IV, No.2, 1944. pg. 17.

7. Smith, L.B. 1960. A Bromeliad Census. Los Angeles, California. The Bromeliad Society Bulletin. Vol. X, pgs. 6-7.

8. Stearn, W.T. 1965. Journal of the Arnold Arboretum. 46:243-285.

9. Proceedings of the Royal Society of London. Vol. XCII. December, 1921, pg. xxvi.

10. Prain, Sir David. 1921. Proceedings of the Linnean Society of London, pg. 41.

11. Padilla, V. Forward. pgs. xiv-xv. Andre, E.F. 1889. *Bromeliaceae Andreanae.* Berkeley, California. Big Bridge Press.

12. Benzing, D.H. 1980. The *Biology of Bromeliads.* Eureka, California. Mad River Press, pg. 4.

13. Padilla, V. 1978. Mulford B. Foster 1888-1978. Los Angeles, California. Journal of the Bromeliad Society, Inc., pg. 243.

14. Smith, L.B. 1969. Personal Recollections of Alberto Castellanos (1896-1968). Los Angeles, California. The Bromeliad Society Bulletin, pg. 124.

Reference Material:

Plumier

Urban, I. 1920. *Reportium specierum novarum regni vegetabilis.* (Fedde). Beihefte 5:1-196.

Linnaeus

Stearn, W.T. 1983. *Botanical Latin.* North Pomfret, Vermont. David & Charles.

Humboldt

Stearn, W.T. 1968. Humboldt, Bonpland, Kunth and Tropical American Botany. Lehre, Germany. J. Cramer.

Schlechtendal & Chamisso.

Eastwood, A. 1944. Schlechtendal & Chamisso. The Botanical Collections of Chamisso and Eschscholtz in California. Leaflets of Western Botany. Vol. IV, No. 2. 1944, pg. 17.

Baker

Watsonia, Journal of the Botanical Society of the British Isles. 1950. Vol. 1, part VI, pgs. 343-344.

Foster

Traub, H.P. 1945, July-October. Biographical Sketches. Plant Life, pg. 99.

Smith

Personal interview with Smith at the Smithsonian Institution in the spring of 1984.

Rauh

Strube, R. 1983. *Professor Dr. Werner Rauh zum 70. Geburtstag.* Frankfurt/Main, Germany. *Die Bromelie. Deutsche Bromeliengellschaft E.V.*, pg. 5.

Chapter 5
Botanical Nomenclature, Pronunciation, & Taxonomy

The Importance of Latin to the Science of Botany

. . . Nobis parienda sunt imponendaque nova rebus novis nomina. Quod quidem nemo mediocriter doctus mirabitur, cogitans in omni arte cuius usus vulgaris communisque non sit multam novitatem nominum esse cum constituantur earum rerum vocabula quae in quaque arte versentur. (Cicero, De Finibus III, 3)

"We are obliged to create a vocabulary and to find names to attach to new discoveries. This will not cause surprise to any moderately well-informed person, when he reflects that in every branch of knowledge lying outside the most elementary ones there must be a large measure of 'newness' about its vocabulary. This 'newness' grows out of the necessity for each area of knowledge to express the particular concepts with which it deals."

The words above by Roman statesman, orator, and man of letters, Marcus Tullius Cicero, were written in the year 45 B.C. Even in those ancient days it was recognized that suitable names for new things and ideas had to be found. Though Cicero believed his native Latin tongue and antecedent Greek would be the main sources for names of new phenomena and concepts, it is doubtful that he could have foreseen the great significance of the role, both direct and indirect, which the two languages have since played in the communication of knowledge.

Botanical Latin originally derives from early Roman writers about plants such as Pliny the Elder (A.D. 23-79). It subsequently survived through the Middle Ages. Latin was then retained during the Renaissance and well into the 18th century as the one internationally used language of learning among the peoples of Europe, none of which at the time possessed a language sufficiently developed and known widely enough to challenge the supremacy of Latin in diplomatic, legal, and ecclesiastical matters.

Thus, during the period when scholars interested in plants were few in every country, their use of Latin helped compensate for their geographical isolation. If their works had been written in their own languages, they would have remained largely unread, unknown, and ineffective.

Although Latin is a "dead" language (it is not used as a mother tongue by any nation), many classical Latin words have been given new definitions over the last few centuries to aid in the description of scientific and/or botanical items or concepts. Additionally, hundreds of new words have been coined to describe the myriad facets of the scientific world. These words have been "latinized" and are therefore, in effect, Latin words whose structure, inflection, and syntax are governed by the rules of classical Latin. Yet the result, as it relates to botany, has been the creation of a new Latin language quite distinct from classical Latin.

To quote the respected British botanist, William T. Stearn (1983), "Botanical Latin is best described as a modern Romance language of special technical application, derived from Renaissance Latin with much plundering of ancient Greek which has evolved, mainly since 1700 and primarily through the work of Carl Linnaeus (1707-78), to serve as an international medium for the scientific naming of plants in all their vast numbers and manifold diversity. The property of no one nation or linguistic group, Latin has, in consequence of its neutrality, become world-wide."

Botanical Taxonomy

Botanical taxonomy, or systematics, is the study of the relationship among plants. To date, more taxonomic work has been done with temperate climate flora than with tropicals. This is because in the world's tropical regions temperatures are warm, it does not freeze, and there is abundant moisture along with high levels of sunlight. As a result, there is a rapid rate of evolution and extinction among plant groups. Species, genera, families, even entire orders develop and disappear quickly within the geological time frame because growing conditions are conducive to rapid plant growth and competition. This makes their classification much more difficult.

In light of this, it is not surprising that no close bromeliad relatives have been discovered, and the fossils, which were first reported in early citations with such optimism, upon closer scrutiny have often turned out to be those of early cycads, pandanas trees, or the faint possibilities that cannot be positively connected to the Bromeliaceae at this time.

As one looks closely at the

evolution and taxonomic structure of plants and animals, it becomes clear that a driving force of evolution is toward the reduction of unnecessary parts or functions. In the Plant Kingdom, the fusion (growing together and uniting) of certain parts facilitates a reduction in the numbers of those parts. This is one reason why monocotyledons with one seedling leaf are considered by and large to be more highly and recently evolved than dicotyledons with two. The dicots are believed to be some millions of years antecedent to the monocots. The difference between dicots and monocots is fundamental in terms of evolutionary development and taxonomically is a class separation.

Dicots, such as magnolias, water lillies, and cacti, have large numbers of stamens and ovaries, with floral parts in fours or fives or multiples thereof, while monocots, such as bromeliads and orchids, have few floral parts—in threes, or multiples thereof.

Among the oldest monocots are palm and pandanas trees which made their appearance some 75-80 million years ago. In contrast, the more recently evolved grasses have exhibited reduction and fusion to an extreme, and represent the cutting edge of plant evolution. The bromeliad family contains species with highly evolved trichomes in the genera of *Tillandsia* and *Vriesea*, but the floral apparatus of species in these genera, and in the family in general, is not as specialized as those of some other families, notably the orchids.

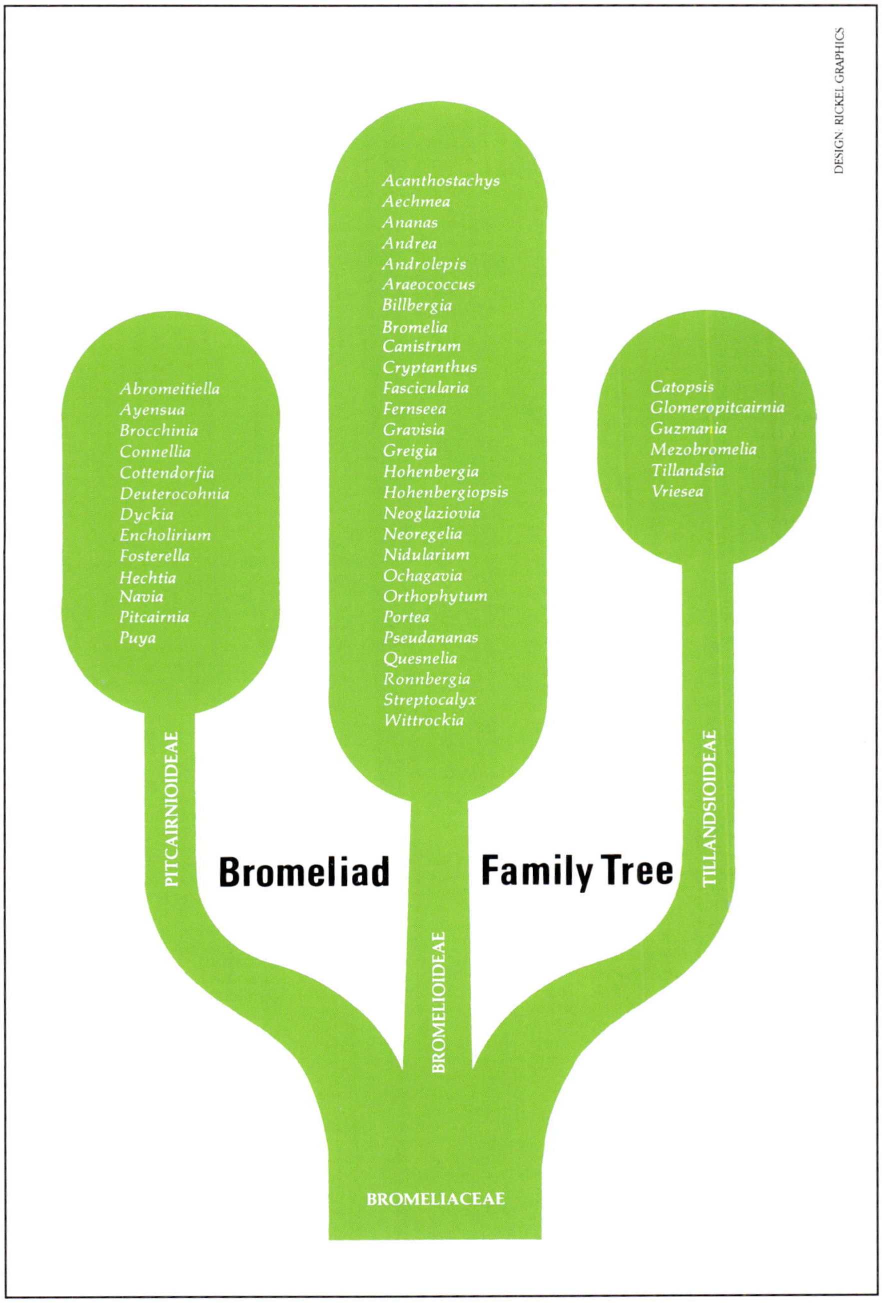

Fig. 5.1

Binomial Nomenclature

All plants, bacteria, and animals on earth are given latinized scientific binomials to individually differentiate species. These named entities are then combined into ever larger related taxonomic groups.

Tillandsia xiphioides v. *tafiensis* from Argentina will be used to illustrate taxonomic classification:

Kingdom	Plantae
Subkingdom	Tracheophyta (includes all vascular plants)
Division	Magnoliophyta (all flowering plants or angiosperms)
Class	Liliopsida (all monocots)
Order	Bromeliales
Family	Bromeliaceae
Subfamily	Tillandsioideae
Genus	*Tillandsia*
Subgenus	*Anoplophytum*
Species	*xiphioides*
Variety	*tafiensis*

The International Code of Botanical Nomenclature (ICBN) regulates all decisions of nomenclature in botany. Strict rules govern the order in which the plants are named and how names are spelled. The current method used to name taxonomic groups and individuals was introduced by the English botanist, John Lindley (1799-1865), in his *Natural System of Botany*, (2nd ed. 1836) in which he made all divisional names of the same value end in the same way.

Divisional epithets end in -phyta, which is originally from the Greek and means PLANT.

Classes end in -opsida, which is from the Greek *opsis*. This translates to "appearance."

Orders end in -ales, which is from the Latin root *-alis* and means "belonging to."

The family name is formed by adding -aceae to the stem of a legitimate name of an included genus, such as Bromeliaceae from the genus name *Bromelia*. A subfamily name is formed in the same manner, with the suffix -oideae being added to a legitimate genus name, such as Tillandsioideae from *Tillandsia*. Both -aceae and -oideae mean a "resemblance."

A generic name is formed by giving the Latin nominative case, feminine singular ending, to the adopted name. The genus, *Tillandsia,* was named by Linnaeus in 1753 after a Finnish botanist, Elias Tillands.

The first letter of a generic name is always capitalized and either underlined or written in *italics*. The species name and varietal name, if there is one, are written entirely in lower case. They must be either underlined or written in *italics*; *Tillandsia xiphioides* v. *tafiensis*. *Xiphioides* is a compound word. *Xiph-* is a Greek base that means "sword," and *-ides* is a Greek suffix which means "like" or "similar to." Together they mean "swordlike" in reference to the shape of the inflorescence.

In written discussion, the generic name is abbreviated unless it begins a sentence or is used for the first time in a paragraph. The generic and species names, and variety if applicable, are usually all that is necessary when referring to a given plant.

A descriptive scientific name often refers to an aspect of the plant—the way it looks, grows, flowers, its habitat—or the plant may be named for the one who first collects it. The discoverer of a plant new to science normally has the honor of naming it if they so choose. The name becomes official when the description is published by the taxonomist.

A natural hybrid is written with the name of the seed parent first and the pollen parent second; a multiplication sign (X) is placed between the names of the parental species, such as *Tillandsia xerographica* X *T. paucifolia*. If a valid description of the hybrid has been published in a scientific work, the plant is given a Latin or latinized name and all future hybrids with the same parental combination will have the same name (collective epithet), which is called the GREX. I described the hybrid above in the *Journal of the Bromeliad Society* and it is known as *T.* X *wisdomiana*. The multiplication sign is placed before the new grex epithet.

Man-made hybrid grexes described after 1959 must be in modern languages, are always capitalized, but are not preceded by an X, nor are they placed in double or single quotes.

Since most of the hybridization in this country has been with *Neoregelia*, and because grex seedlings of this genus are so variable in size and coloring (creating a high demand for new cultivars), it will be used to illustrate the system. A hypothetical example would be a cross between two species such as *Neo. concentrica* L.B. Smith X *Neo. compacta* L.B. Smith. The plants of the resulting hybrid grex could then be given an epithet such as *Neoregelia* Brazilian Pride.

A CULTIVAR, which is short for "cultivated variety," is a plant within a grex that is markedly different from the norm and reproduces, either from seed or vegetatively or both, true to that form—that is, maintaining the distinguishing characteristic(s). Up to three descriptive modern language words can be used in a cultivar name.

I have used the cultivar format to give names to subspecies from wild, such as *Tillandsia disticha* 'Major.' These plants are not descended from a single genetic clone and are not cultivars. They are actually subspecies but until such time as they are properly described in Latin in a valid publication, they should not be referred to in print with a subspecific epithet, such as *T. disticha* v. *maxima*.

The name of a cultivar can be written in either of two ways: by placing cv. before the epithet, as with *Achmea orlandiana* cv. Ensign, or by enclosing the name within single quotes, such as *Achmea orlandiana* 'Ensign.' This second method is normally reserved for clonal designation. Cultivars are always capitalized, and since 1959, they must be in modern languages—not Latin. They are not underlined or *italicized* in texts.

A clone is a group of genetically identical individuals descended from a single parent by vegetative, asexual (mitotic) reproduction. Clones cannot be reproduced from seed as can cultivars.

A cultivar name can be used to differentiate one or more superior plants from the same parental seedling batch. The future vegetative offsets of that cultivar would also be cultivars and genetic clones of the original plant.

In bromeliads at least, F_1 (self-pollinated) seedling descendants of a hybrid are not normally cultivars. The reason is that one of the cultivar criteria, sexual reproduction true to form, does not hold. F_1 plants are noted for their variation and unless the entire batch of plants is virtually identical to the parent, they will not carry the parental cultivar name.

In our hypothetical example, a hybrid batch of *Neoregelia* was given the grex name of *Neoregelia* Brazilian Pride. A superior plant from the grex might then have been given a cultivar name such as *Neoregelia* Brazilian Pride cv. Betty's Bonanza. Its future pups through the generations would be clones as well as cultivars, and the epithets would be written *Neo.* Brazilian Pride 'Betty's Bonanza.' However, plants of any future F_1 seedling batches of *Neo.* Brazilian Pride cv. Betty's Bonanza would be neither cultivars nor clones, but would be part of the original grex of *Neoregelia* Brazilian Pride. Exemplary plants from the grex could then be given new cultivar names.

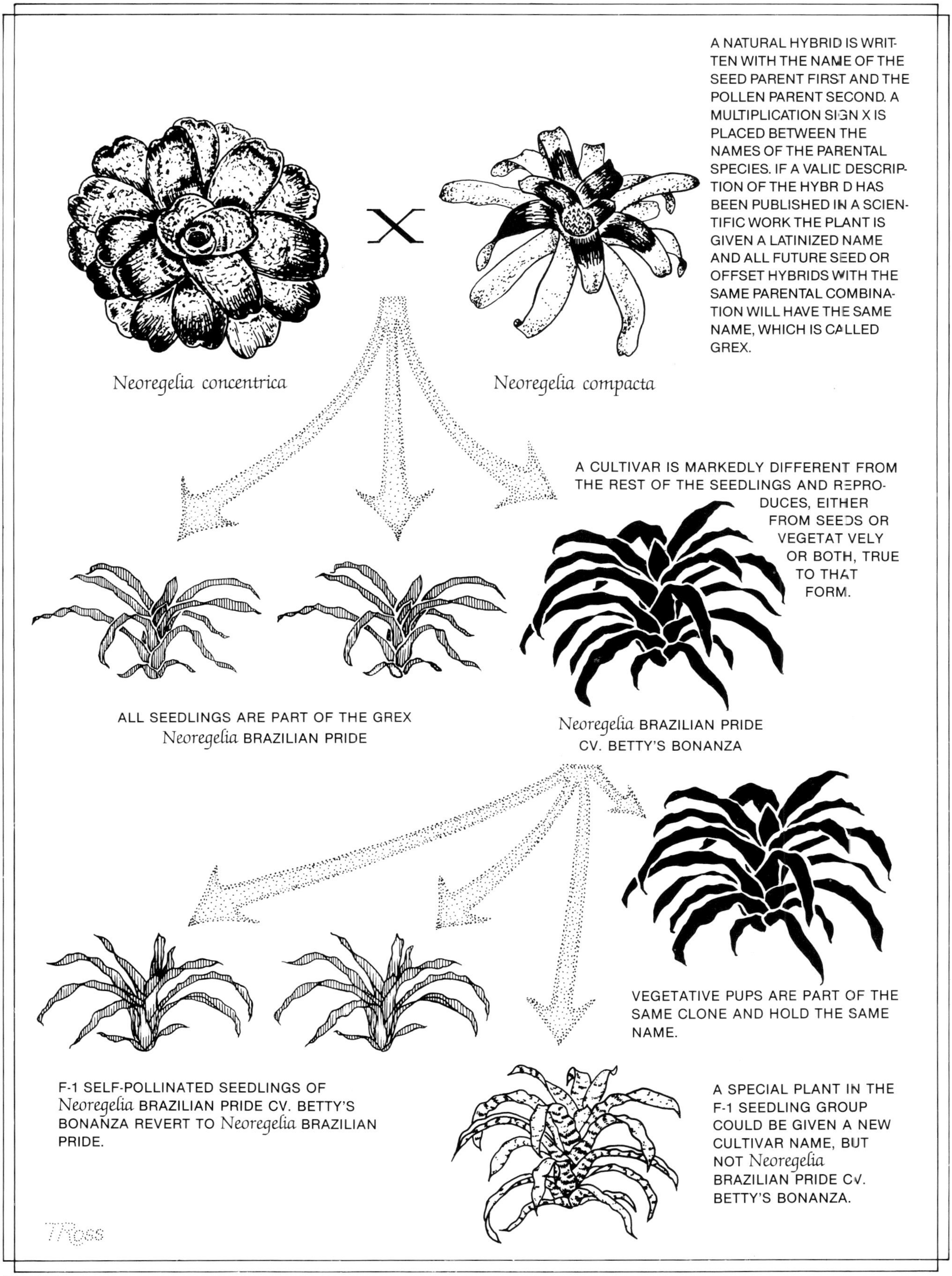

A NATURAL HYBRID IS WRITTEN WITH THE NAME OF THE SEED PARENT FIRST AND THE POLLEN PARENT SECOND. A MULTIPLICATION SIGN X IS PLACED BETWEEN THE NAMES OF THE PARENTAL SPECIES. IF A VALID DESCRIPTION OF THE HYBRID HAS BEEN PUBLISHED IN A SCIENTIFIC WORK THE PLANT IS GIVEN A LATINIZED NAME AND ALL FUTURE SEED OR OFFSET HYBRIDS WITH THE SAME PARENTAL COMBINATION WILL HAVE THE SAME NAME, WHICH IS CALLED GREX.
X
Neoregelia concentrica
Neoregelia compacta
A CULTIVAR IS MARKEDLY DIFFERENT FROM THE REST OF THE SEEDLINGS AND REPRODUCES, EITHER FROM SEEDS OR VEGETATIVELY OR BOTH, TRUE TO THAT FORM.
ALL SEEDLINGS ARE PART OF THE GREX *Neoregelia* BRAZILIAN PRIDE
Neoregelia BRAZILIAN PRIDE CV. BETTY'S BONANZA
VEGETATIVE PUPS ARE PART OF THE SAME CLONE AND HOLD THE SAME NAME.
F-1 SELF-POLLINATED SEEDLINGS OF *Neoregelia* BRAZILIAN PRIDE CV. BETTY'S BONANZA REVERT TO *Neoregelia* BRAZILIAN PRIDE.
A SPECIAL PLANT IN THE F-1 SEEDLING GROUP COULD BE GIVEN A NEW CULTIVAR NAME, BUT NOT *Neoregelia* BRAZILIAN PRIDE CV. BETTY'S BONANZA.
T.Ross

Fig. 5.2

Species Pronunciation

There are two systems in use by which one may accomplish accuracy and consistency in the pronunciation of species' names. One is to pronounce the words with a reasonable approximation of Latin as it was spoken by educated Romans and is still spoken today by Europeans and Latin Americans. The other is to use the Anglicized (English accent or traditional English) method that is in general use among English speaking botanists and gardeners.

After studying the two systems, I realized that both of them caused changes in the way I had previously spoken the names of some species. In the end I decided to use Latin pronunciation for a number of reasons. First, Latin is correct and universally accepted. And importantly, the rules of pronunciation in Latin are precise and consistent, unlike English. Once understood, they can be used for all words, which makes learning them much easier. English has many words for which the pronunciation must be memorized because the letters and/or syllables are "exceptions to the rule." As anyone who has had to learn English as a second language can appreciate, this is very difficult.

Also, the species' names are Latin, not English. Latin was used to create a system by which persons of differing tongues could communicate with each other. No one country had the right to say Latin should be spoken with their accent. Anglicized Latin is not well understood or received outside English speaking countries.

The plants presented in this book have the Latin pronunciations listed just below their names. Many Americans will want to pronounce familiar-looking words in standard American dialect. For instance, *tricolor* will be spoken as TRY-col-or, not TREE-col-or, and *ionantha* as eye-oh-NAN-tha instead of ee-oh-NAHN-tha, and that is fine. The pronunciation guide is primarily intended for more unfamiliar looking words, but the rules have been used consistently for all species.

PRONUNCIATION GUIDE*

Reformed Academic	Traditional English
ā as in *fāther*	*fāte*
ă as in *ăpart*	*făt*
ae as *ai* in *aisle*	as *ea* in *meat*
au as in *house*	as *aw* in *bawl*
c always as in *cat*	before *a, o, u* as in *cat* before *e, i, y* as in *center*
ch (of Greek words) as *k* or *k-h* (if possible).	as *k* or *ch*
ē as in *they*	*mē*
ĕ as in *pĕt*	*pĕt*
ei as in *rein*	as in *height*
g always as in *go*	hard before *a, o, u,* as in *gap, go* soft before *e, i, y* as in *gem, giro*
ī as in *machīne*	*īce*
ĭ as in *pĭt*	*pĭt*
(consonant i) as *y* in *yellow*	j in *jam*
ng as in *finger*	*finger*
ō as in *nōte*	*nōte*
ŏ as in *nŏt*	*nŏt*
oe as *oi* in *toil*	as *ee* in *bee*
ph as *p* or *p-h* if possible	like *f*
r always trilled	
s as in *sit, gas*	*sit, gas*
t as in *table, native*	*table* but *ti* within a word as in *nation*
ū as in *brūte*	*brūte*
ŭ as in *fŭll*	*tŭb*
ui as *oui* (French), *we*	*ruin*
v (consonant *u*) as *w*	as in *van*
ȳ as *u* in French *pur*	as in *cȳpher*
y̆ as in French *du*	as in *cy̆nical*

*Reprinted from: Stearn, W. T. 1983. *Botanical Latin.* Permission granted by publishers: David & Charles, Brunel House, Newton Abbot, Devon, England.

Fig. 5.3

Species named after individuals are pronounced as that person said it. Therefore, *Tillandsia rothii* would be spoken as ROTH-ee-ee, and not ROTE-ee-ee. Some species names have their origins in ancient Greek. These words have been transliterated into Latin by taxonomists.

I have avoided some of the minor differences between Latin and English in trying to balance correct Latin pronunciation and keeping the changes simple and basic. For instance, in Latin, the letter "s" is pronounced as the letter "z" when it is located between two vowels. The species name *plumosa*, ploo-MOE-za, is actually more correct than ploo-MOE-sa. Also, I have overlooked a few Latin sounds that are foreign to the American ear. For instance, in classical Latin, the letters "g" and "c" are always hard, as in *go* and *cat*. The long "y" sounds as the "u" in the French *pur*, and the short "y" is said as the "u" in the French *du*. The letter "v" is spoken as an English "w."

Finally, I have used Latin pronunciation for species, genus, and family names only. When given, the pronunciation of botanical words is in standard American dialect.

Although not complete, here are a few rules of Latin pronunciation. With words of two syllables, the accent is always on the first syllable, i.e. CA-put, not ca-PUT. For words of more than two syllables, the accent is on the second-to-last syllable (penultimate) if the vowel in that syllable is long, as in the species name *fas-ci-cu-LA-ta* (fah-sih-cuh-LAH-ta). The accent is on the third-to-last syllable (antepenultimate) if the vowel in the penultimate syllable is short, that is for the name straminea (strah-MI-nay-a). In Latin or latinized Greek, every vowel is pronounced, hence, ix-i-o-i-des (ix-ee-oh-EE-des), not ix-i-oi-des (ix-ee-OY-des).

Tillandsia Taxonomy

This presentation is based on the criteria used by Mez and Smith to classify the plants. The reader should be aware that it is presented more in the vein of a historical starting point—a still photograph from a constantly playing motion picture. This discussion is not, and could not be, a "state of the art" treatise. Some of the taxonomic material in the book will be dated by the time it is published.

With curiosity and perseverance a person may easily learn about *Tillandsia* in greater depth. The botanical terms and taxonomic relationships are not complicated, but they do require some study. Once one is familiar with the more technical words, their meanings will stay. Use of the glossary at the end of this book will help greatly. There is inherent satisfaction once these words are understood and used in communication, for they are the "tools" to ask more and better questions.

Traditionally, systematists have described species from dried herbarium specimens. Vegetative characteristics have long been used for taxonomic classification, because many parts of the floral apparatus, such as the flower petals, stamens, and pistils, do not preserve well when dried.

Increasingly, however, floral traits are being used to classify angiosperms. The floral structure is usually the slowest part of the plant to evolve (physically change) in response to natural selection and is, therefore, the best part of the plant to use in systematics. The opportunity to preserve fresh floral organs and the development of the scanning electron microscope (SEM), along with other new technology, is now providing taxonomists with new opportunities to more accurately classify plants. In addition, botanists are increasing the use of computer-aided analyses in their efforts to classify and estimate evolutionary pathways. Taxonomic revision of Tillandsia has begun (Gardner, 1982; Gilmartin & Brown, 1984, 1986).

All bromeliads are herbs. An herb is a plant that does not have a permanent woody stem above ground. The bromeliad family is divided into three subfamilies: the Pitcairnioideae (pit-care-knee-oh-EE-day-eye), the Bromelioideae (bro-mehl-ee-oh-EE-day-eye), and the Tillandsioideae (til-lands-ee-oh-EE-day-eye).

One way to distinguish between genera in Tillandsioideae and those of the other two subfamilies is that the leaves of all species of Tillandsioideae are entire—they never have spines along the edges of the leaves. The leaf margins of species in the other two subfamilies usually possess spines.

Tillandsioideae genera include the following: *Tillandsia* Linnaeus, with some 500 currently recognized species; *Vriesea* Lindley, with approximately 250 species and many man-made hybrids; *Guzmania* Ruiz & Pavon, with about 126 species and numerous man-made hybrids; *Mezobromelia* L.B. Smith, with only three species; *Catopsis* Grisebach, with about 19 species; and *Glomeropitcairnia* Mez, with just two species.

All species in subfamily Tillandsioideae have an indumentum of radially symmetric scales. They are sparsely distributed in mesic *Tillandsia*, most *Vriesea*, and all *Guzmania*, *Catopsis*, and *Glomeropitcairnia*. Trichomes are dense in xeric species of *Tillandsia* and a few in *Vriesea*. One cannot be sure which genus is home for an unfamiliar plant without a look at the bloom because of the generic overlap in characteristics such as shape, size, color, and habitat preference.

Taxonomic Division of Tillandsioideae Genera

In all Tillansioideae genera, except *Glomeropitcairnia*, the ovary is superior, that is the sepals, petals, and stamen filaments are attached to the receptacle as discrete organs below the base of the pistil (and ovary). This more primitive arrangement of the floral structure is termed HYPOGYNOUS. *Glomeropitcairnia* is the only genus that produces seeds that have equal-size plumose appendages (fine, feather-like hairs) at both ends of the seed. These appendages are an outgrowth of the seed coat.

The hairs of all *Tillandsia* are straight at maturity and shaped like a parachute. The hairs of *Catopsis* seeds are folded and wavy at maturity. The sheaths of *Catopsis* are often glaucous.

The inflorescences of *Guzmania* and *Mezobromelia* are always polystichous and the petals are conglutinated (con-glue-ti-nay-ted), which means they are firmly united together in a tube. *Guzmania* and *Mezobromelia* are separated from each other because the petal-claws of *Mezobromelia* bear LIGULAE on the adaxial surfaces.

The petals of *Tillandsia* and *Vriesea* are FREE (unattached), and the flower spikes are often distichous (two-ranked).

Baker included *Vriesea* as a subgenus of *Tillandsia* in 1889, but Mez changed this and the practice was continued by Smith. They used the presence or absence of two ligulae on the adaxial base of flower petals to separate Tillandsia and Vriesea into two distinct genera.

These flaps, according to bromeliad biologists Utley (1978), Benzing (1980), and Gardner (1982) are of ambiguous biological significance. Benzing states, "Perhaps the most arbitrary taxonomic boundary standing in *Tillandsioideae* today is that between members of its two largest genera. . . .If petal flaps are present, the individual belongs to *Vriesea*; if not, you have a *Tillandsia*. Petal scales are of equivocal biological significance, consisting of nothing more than a pair of delicate flaps of tissue arising from the inner surface of each corolla member. There are tillandsias and vrieseas that, except for these petal claws, are much better suited to be in the other genus."

Indications now point to a restructuring of *Tillandsia* subgenera as more scientific knowledge comes to light.

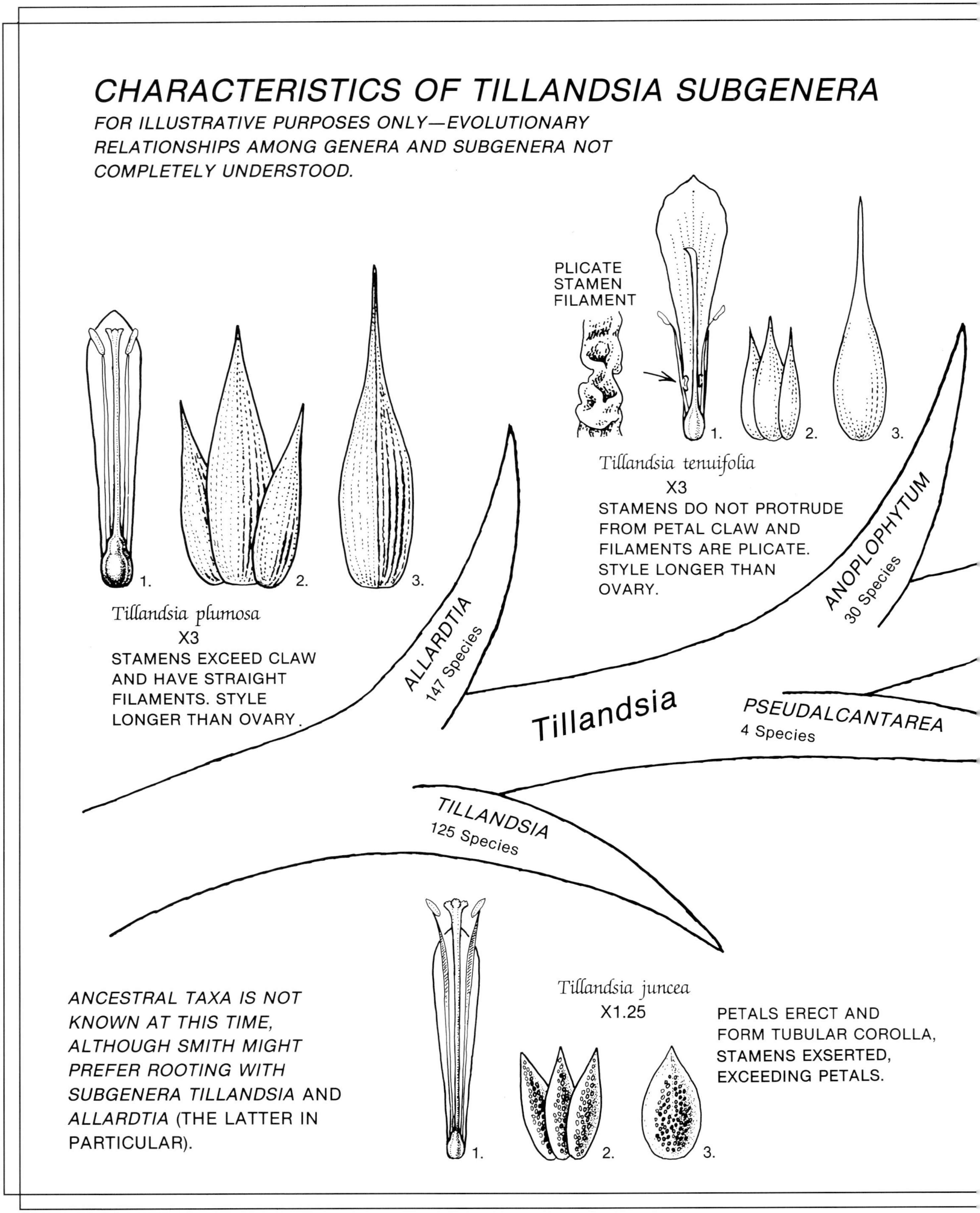

CHARACTERISTICS OF TILLANDSIA SUBGENERA
FOR ILLUSTRATIVE PURPOSES ONLY—EVOLUTIONARY RELATIONSHIPS AMONG GENERA AND SUBGENERA NOT COMPLETELY UNDERSTOOD.
PLICATE STAMEN FILAMENT
1.
2.
3.
Tillandsia tenuifolia
X3
STAMENS DO NOT PROTRUDE FROM PETAL CLAW AND FILAMENTS ARE PLICATE. STYLE LONGER THAN OVARY.
Tillandsia plumosa
X3
STAMENS EXCEED CLAW AND HAVE STRAIGHT FILAMENTS. STYLE LONGER THAN OVARY.
ALLARDTIA
147 Species
ANOPLOPHYTUM
30 Species
Tillandsia
PSEUDALCANTAREA
4 Species
TILLANDSIA
125 Species
ANCESTRAL TAXA IS NOT KNOWN AT THIS TIME, ALTHOUGH SMITH MIGHT PREFER ROOTING WITH SUBGENERA TILLANDSIA AND ALLARDTIA (THE LATTER IN PARTICULAR).
Tillandsia juncea
X1.25
PETALS ERECT AND FORM TUBULAR COROLLA, STAMENS EXSERTED, EXCEEDING PETALS.

Fig. 5.4

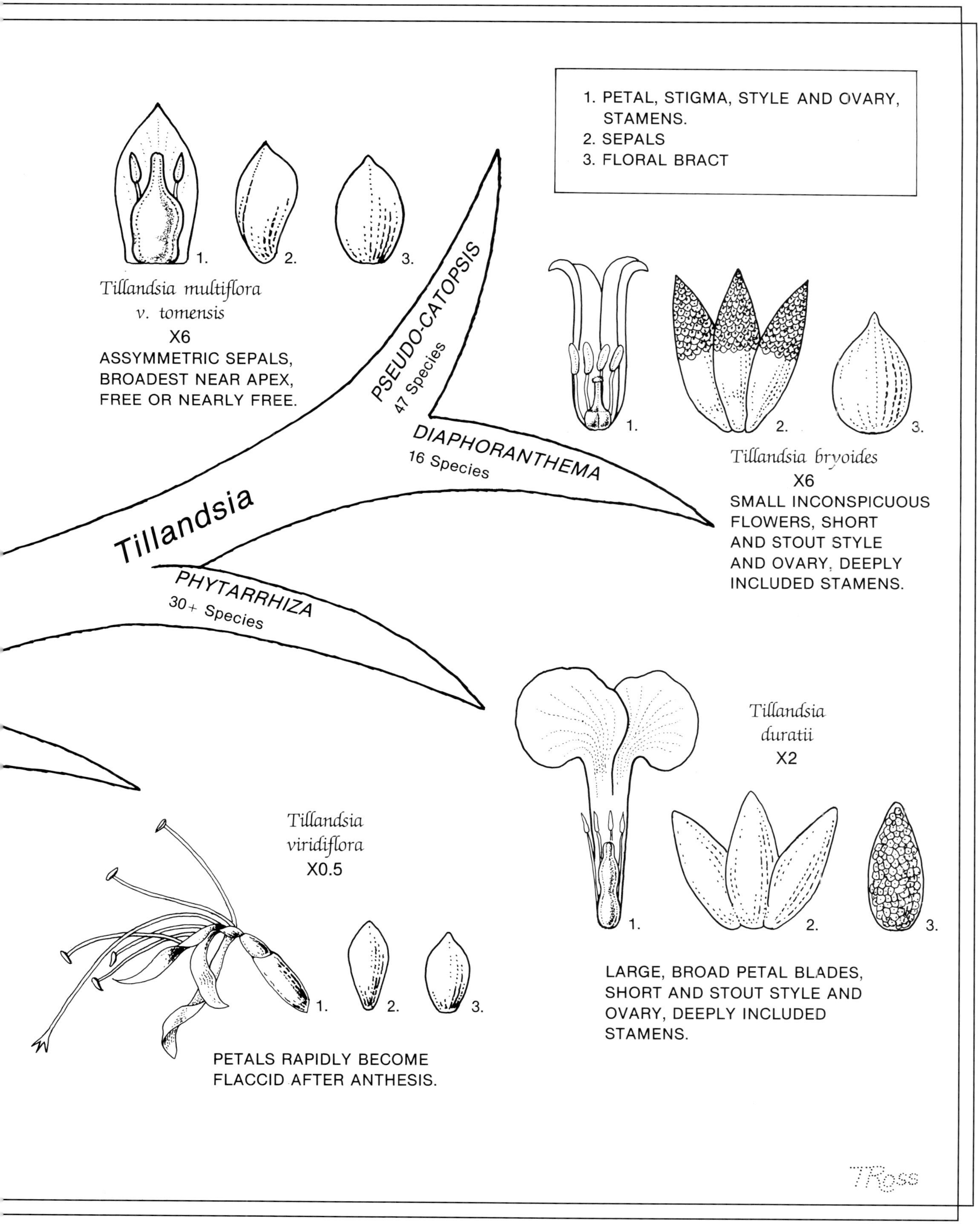
1. PETAL, STIGMA, STYLE AND OVARY, STAMENS.
2. SEPALS
3. FLORAL BRACT
1.
2.
3.
Tillandsia multiflora v. tomensis
X6
ASSYMMETRIC SEPALS, BROADEST NEAR APEX, FREE OR NEARLY FREE.
PSEUDO-CATOPSIS
47 Species
DIAPHORANTHEMA
16 Species
Tillandsia
PHYTARRHIZA
30+ Species
1.
2.
3.
Tillandsia bryoides
X6
SMALL INCONSPICUOUS FLOWERS, SHORT AND STOUT STYLE AND OVARY, DEEPLY INCLUDED STAMENS.
Tillandsia duratii
X2
1.
2.
3.
LARGE, BROAD PETAL BLADES, SHORT AND STOUT STYLE AND OVARY, DEEPLY INCLUDED STAMENS.
Tillandsia viridiflora
X0.5
1.
2.
3.
PETALS RAPIDLY BECOME FLACCID AFTER ANTHESIS.
TRoss

Taxonomic Division of Genus *Tillandsia* Subgenera

Presently, there are seven described *Tillandsia* subgenera. Some have just a few species and others contain more than a hundred. To differentiate them, one must dissect a flower.

The first subgenus to be separated is *Pseudo-Catopsis* Baker. In contrast to members of the other subgenera, the sepals are asymmetric, broadest near the apex, and free or nearly free. The inflorescences are typically tall, with many lax, spreading, branching spikes that look like those of species in genus *Catopsis*. The type species is *Tillandsia ropolacarpa* Andre from Ecuador. *Pseudo-Catopsis* contains close to 50 species, but only a few are in cultivation. *Tillandsia fraseri* Baker, a green ligulate-leaved, tank-type species with a magnificent, cherry red inflorescence, is the only species commonly available at this time.

The second subgeneric differentiation is whether the stamens are exserted above or included in the throat of the corolla. If they are exserted, the plant has to be in either subgenus *Tillandsia* Linnaeus or *Pseudalcantarea* Mez (sue-dahl-conTAR-ay-a). One difference between these two subgenera is that the petals of *Tillandsia* species are erect, that is, the three petals make up a long, tubular corolla. In addition, the leaves of most species are narrowly triangular. The petals of species in *Pseudalcantarea* are subspreading and become rapidly flaccid (soft and weak). All four of the large species in this subgenus are mesic and form rosettes. The type is *Tillandsia viridiflora* Beer. Two species, *T. baliophylla* Harms and *T. paniculata* Linnaeus, are restricted to Hispaniola in their distribution. The other two species, *T. viridiflora* and the large growing *T. grandis* Schlechtendal, grow from Mexico to Nicaragua in Central America. In bloom, *Tillandsia paniculata* can reach a height of four and a half meters.

Subgenus *Tillandsia* currently has over 115 recognized species ranging from Georgia and Florida in the north, through Mexico, Central America, and into South America. Species occur as far south as southern Brazil. *Tillandsia utriculata* Linnaeus is not only the type for the subgenus but the genus as well. Many of the most popular, widely available, and easy-to-grow species are contained in *Tillandsia*.

Tweezers and a loop or magnifying glass are now used to take a flower apart. The flower can be cut off at its point of attachment to the receptacle. The petals and sepals should be carefully bent back and cut away, exposing the ovary, stigma, and stamens.

The style is the tube that goes up from the ovary to the stigma. One differentiation between subgenera is whether or not the style is longer than the ovary and slender, or short and stout. Another is whether the stamens are longer or shorter than the length of the petal claw (the stalk or tubelike part of the corolla).

If the style is longer than the ovary and slender, and the stamens are at least equal to the length of the petal claw, the plant is in either *Allardtia* (a-LARD-tee-a) Baker or *Anoplophytum* (ah-no-plo-FIH-tum) Baker. If the style and stamens are short and stout, it is in either *Phytarrhiza* (fih-tar-EE-za) Baker or *Diaphoranthema* (dee-a-for-AHN-tha-ma) Baker.

To separate *Allardtia* and *Anoplophytum*, one must look at the stamens. If they exceed the claw of the petal and have straight filaments, the plant is in *Allardtia*. This taxon currently has close to 150 recognized species, the most of any subgenus. The type is *T. guatemalensis* L.B. Smith. Species in *Allardtia* range from Mexico, Central America, and the Caribbean to South America.

Unlike the members of *Allardtia*, the filaments of species in *Anoplophytum* are not straight but often strongly plicate (folded like a fan). *Anoplophytum*, with just over 30 described species, has stamens that are about equal in length to the petal claw. The type species is *Tillandsia stricta* Solander.

As previously stated, the last two subgenera were separated from *Allardtia* and *Anoplophytum* by a short, stout style along with deeply included stamens. Species in *Phytarrhiza* have large, broad, easily visible petal blades. The subgenus has over 30 species. The type is *Tillandsia duratii* Visiani. Species range from Nicaragua in Central America through South America. Many of the most spectacularly shaped and fragrant species are in this subgenus.

Species in *Diaphoranthema* have small, inconspicuous flowers. The type is *Tillandsia recurvata* Linnaeus. The range of *Diaphoranthema* is the greatest of any subgenus—from the southern United States to southern Argentina and Chile. However, it has a low cohesion ratio.* In this case, only a small area of Argentina and Chile is common to all species. It has been suggested by a number of people that its widespread distribution is due to long distance dispersal by birds (Benzing 1980, McWilliams 1974). *Diaphoranthema* contains most of the minature species, including the ubiquitous *T. usneoides*, or "Spanish Moss."

Literature Cited:

Benzing, D.H. 1980. *The Biology of Bromeliads*. Eureka, California. Mad River Press, pg. 147.

Brown, G. & Gilmartin, A.J. 1984. Stigma structure and variation in Bromeliaceae—neglected taxonomic characters. Brittonia, 36(4). pgs. 364-374.

Gardner, C.S. 1982. *A Systematic Study of Tillandsia Subgenus Tillandsia*. Ph.d. dissertation. College Station. Texas A & M University.

Gardner, C.S. 1982, pg. 2.

Gilmartin, A.J. & Brown, G. 1986. Cladistic tests of hypotheses concerning evolution of xerophytes and mesophytes within *Tillandsia* subgenus *Phytarrhiza* (Bromeliaceae). Amer. Journal of Botany. Vol. 73, No.l 3. pgs. 387-397.

McWilliams, E.L. 1974. Evolutionary Ecology in *Flora Neotropica Monograph No. 14*. pg. 48.

Stearn, W.T. 1983. *Botanical Latin*. North Pomfret, Vermont. David & Charles, pgs. 6-7.

Utley, J.F. 1978. A revision of the Costa Rican Thecophylloid Vrieseas (Bromeliaceae). Ph.D. dissertation. Duke University.

Reference Material:

Handbook for *Judges and Exhibitors*. 1982. Austin, Texas. Bromeliad Society, Inc.

*The term used by L.B. Smith (1934) to describe a situation in which a much smaller region relative to total distribution includes representatives of all species.

Chapter 6
Tillandsia Evolution & Biology

Tillandsia Evolution and Morphology

One point that both fascinates and confounds those who have an interest in *Tillandsia* is the fantastic array of shapes, colors, and sizes exhibited by members of this amazing genus. That so many closely related species could be so visibly different in so many ways sometimes strains the limits of our credulity. We are ecstatic as the multitude of *Tillandsia* characteristics and variations delight our senses and challenge our intellects. We are thrilled when they bloom with parrotlike displays of brilliant color, delighted when they reproduce abundantly. When we have grown tillandsias into large clumps that have been with us through thick and thin, season after season, they become dear indeed.

Tillandsias are relatively recent additions to the world's flora. They, like all living plants, have evolved from extinct groups. At some fairly recent point in the past, probably some tens of millions of years ago, early bromeliad ancestors began to diversify into distinct taxonomic lines. Why? Because that is the nature of what Charles Darwin termed NATURAL SELECTION. Natural selection is the *cause*, not the *product* of evolutionary radiation.

"Again, it may be asked, how is it that varieties, which I have called incipient species, become ultimately converted into good and distinct species which in most cases obviously differ from each other far more than do the varieties of the same species? How do those groups of species, which

Fig. 6.1 Charles Darwin

constitute what are called distinct genera, and which differ from each other more than do the species of the same genus, arise? All these results . . .follow from the struggle for life. Owing to this struggle, variations, however slight and from whatever cause proceeding, if they be in any degree profitable to the individuals of a species, in their infinitely complex relations to other organic beings and to their physical conditions of life, will tend to the preservation of such individuals, and will generally be inherited by the offspring. The offspring, also, will thus have a better chance of surviving, for, of the many individuals of any species which are periodically born, but a small number can survive. I have called this principle, by which each slight variation, if useful, is preserved, by the term Natural Selection, in order to mark its relation to man's power of selection."

So stated Darwin, the Father of Evolutionary Theory in his classic work, *On the Origin of Species,* first published in England in November 1859. Natural selection is the process by which living organisms either adapt to changing physical and biotic environmental conditions or eventually become extinct. Natural selection acts on the entire plant during all stages of growth, from seed to flower, and back to seed. In order to understand natural selection, one must first look at the basic genetic nature of the plants—and it is in this area that Darwin's theory contained a major flaw.

Darwin knew little about the causes of inherited variation, and his opinions on the subject were neither logical nor consistent. He regarded hereditary substances as fluid in nature rather than particulate, and the observed intermediate nature of hybrids between races or breeds as resulting from a mixture of parental fluids in their bodies. The evidence that heredity is determined not by fluids but by discrete, submicroscopic, units called GENES already existed in the records of Gregor Mendel, an Austrian monk who lived in Brunn, Czechoslovakia. Darwin knew nothing of this.

Natural Selection As a Tool for Phylogenetic Change

G. Ledyard Stebbins (1966) lists several ways in which natural selection acts to bring about differences among species. They are interesting but too involved to be discussed in much detail. However, two of the methods may give a clue to some of the variation among *Tillandsia* species.

Described by Ernst Mayr in 1963 as the FOUNDER PRINCIPLE, the first seed or seeds of a species in a new geographical area are founders. The founder theory deals with the effects on the ensuing population of the particular combinations(s) of characters held by the founder(s). For instance, if a founder plant was an atypical (varied) member of a species, the subsequent population to evolve from that plant would already be well on its way toward a new and different genotype-phenotype*—that is, species or variety thereof.

Tillandsia founder populations developing from wind- or bird-dispersed seeds consist of one or very few plants at first. If successful, the plant colony will expand until a state of equilibrium is achieved with the selective pressures of the environment.

A factor that may influence the differentiation of some populations is CHANCE. When a wind-dispersed *Tillandsia* population begins to establish itself, genetic characters that control relatively LOW ADAPTIVE VALUES may oscillate in frequency. Low adaptive values are those that have no significant bearing on the NEUTRALITY (neither adaptive or nonadaptive in so far as we know) of a plant. An example of a low adaptive value in one case might be leaf length, width, or straightness.

Phylogenetic change could begin in one generation, if there is an exceptionally high proportion of (chance) fertilizations between plants that are homozygous (AA or aa) for one of a pair of ALLELES. This would temporarily increase the frequency of that allele in the population.

In a small, isolated population, such as would occur with a newly established *Tillandsia* colony, such a fluctuation could lead to the complete loss of the opposite allele from the population. The first chance mating of the two homozygous alleles would become FIXED and a permanent part of the genotype. This interaction between selection in a founder population and chance is probably responsible for many of the morphological differences among *Tillandsia* species.

But as Sewall Wright (1982 and 1983) has demonstrated, chance elimination of a gene, which is known as GENETIC DRIFT, will not occur in a large population. A

Fig. 6.2 Gregor Mendel

deviation in one direction on the part of a few individual plants is certain to be balanced by a chance deviation in the opposite direction elsewhere in the population. This is why Mayr and others believe that speciation can occur only in small, genetically (reproductively) isolated populations.

Moreover, chance differences such as leaf length and straightness may determine ways species adapt to similar environments. These will become accentuated as the environment changes further. Chance combinations of genetic characters with low adaptive values by themselves probably have little effect on evolutionary processes. But in combination with the small numbers in a new population, and the accompanying natural selection, they may play an important role in guiding the early stages of adaptive differentiation.

In addition, nonadaptive or neutral differences among species might arise because an adaptive gene may direct one or more other nonadaptive traits. When a gene directs more than one character trait, it is said to have a PLEIOTROPIC effect. If a nonadaptive, pleiotropic effect occurs early in the establishment of the population, it too can become a permanent part of the species genotype.

For instance, the sickling of red blood corpuscles (sickle cell anemia) in some human populations is produced by the heterozygous, Ss, condition for a gene that is lethal in the homozygous, ss, condition. People having the heterozygous trait, Ss, are mildly anemic and so would seem to be at a disadvantgage. Yet this trait can reach a frequency of 40% in some African black populations. The reason is that heterozygotes for the sickling gene are more resistant to malaria than are homozygous normal, SS, individuals. Therefore, the red blood cell behavior that causes the anemia is inconsequential to a much more important physiological difference—resistance versus susceptibility to malaria. As one would expect, differences in rate of occurrence of sickle cell anemia between populations in which malaria is present and those that are free of the disease have been found (Stebbins; 1966).

Phylogenetic Change Through Hybridization

Often, two or more species grow together in the same habitat, and when this occurs among closely related taxa there is opportunity for HYBRIDIZATION. In most cases of hybridization there is a single individual plant, or occasionally a few individuals, that are intermediate between the two parental species growing in the same area. This type of hybridization is fairly common in *Tillandsia* (Smith & Downs, 1977; The Bromeliad Society, Inc., 1977; Gardner, 1984). Hybrids will be genetically different from either parental species and will probably be genetically superior to them in some way, or the forces of natural selection will see to it that they do not survive. Indeed, this is most often the case and probably the reason that most PUTATIVE (probable, not scientifically proven)

*A genotype is the genetic constitution of an individual as opposed to its appearance, or phenotype.

hybrids are isolated, few in number, and never develop into larger populations.

Stebbins (1966) has shown that when a hybrid is successful, there is a greater probability that the new population will become established, not by the mating of the hybrids among themselves, but rather by the mating of original hybrids (called F_1 hybrids) back to one of the parental species, that is BACK CROSSING. This will be followed by a gradual selection of genotypes having a preponderance of genes derived from one species, but importantly, including a few chromosomal segments introduced from the other species, a key for genetic variability.

There are three reasons why F_1 hybrids are more likely to form progeny by back crossing to one of the parental species than by mating with each other. First, at least initially, the hybrids are fewer in number than are individuals of the parental species. Second, the high fertility of parental plants as compared to the low fertility of hybrids will make an enormously greater number of fertile flowers available for pollinating the hybrids. Third, back cross progeny, since they may contain genes derived principally from one of the parental species, are the most likely to be well adapted to a habitat already present, although plumose seed dispersed far distances may end up in environments that are quite different, as previously discussed.

This sequence of events—hybridization, back crossing, and stabilization of back cross types into new species by the process of natural selection is known as INTROGRESSION. Introgression is a necessary condition for the establishment of the segregating progeny of a hybrid population. This population is termed a HYBRID SWARM (Anderson, 1949). The significance of a hybrid swarm is that it may provide a rapid increase in the size of the gene pool from which natural selection for adaptation to a new habitat might quickly sort out entirely new adaptive gene complexes. The hybrids are usually vigorous and intermediate in most respects between the parental species. When seeds are grown from such hybrids, they produce a large and extremely variable segregating progeny, no two individuals of which are genetically alike. Furthermore, when such progeny are grown in a variety of environments, individuals can be recognized as having properties different from the original parental species, as well as from the F_1 hybrid.

Fig. 6.3 *Tillandsia jucunda* Castellanos.

After some generations, the population will become a new species as it develops a stable genotype. Gardner (1984) suggests that *Tillandsia polita* L.B. Smith is a putative hybrid between *T. fasciculata* v. *rotundata* L.B. Smith and *T. rodrigueziana* Mez. Likewise, Gardner states that *T. rothii* Rauh may be a hybrid of *T. roland-gosselinii* Mez and *T. jaliscomonticola* Matuda. My own observation is that *T. jucunda* Castellanos is a putative hybrid between the sympatric *T. ixioides* and *T. tenuifolia*.

The introduction of new hybrids in a range can move evolution forward much more rapidly than through the gradual accumulation of new genetic traits passed on to each successive generation. In this manner, entire populations of the new genotype can be established with relative speed.

POLYPLOIDY is a process sometimes associated with hybridization. It is a method by which chromosome sets are multiplied. Polyploidy is not common in bromeliads. With the exception of most species in subgenus *Diaphoranthema*, which have a diploid number, N, of 100 chromosomes (N=50, 2N=100 pairs), all *Tillandsia* species studied to date have a chromosome count of N=25 or less (Brown 1985; personal communication). According to Brown, there is a definite trend toward polyploidy within all of *Diaphoranthema*, the subgenus some authorities consider the most highly evolved in the genus. Because polyploid species of *Diaphoranthema* have double the normal count of 25 chromosomes, they are termed TETRAPLOID.

Stebbins (1971) has estimated that polyploidy has played a significant role in the evolution of 40-50% of the angiosperm species on earth. This is because polyploidy has helped stabilize hybrids of widely disparate genomes. As a diploid, a typical interspecific hybrid is probably sterile to some degree, because the chromosomes do not have sufficient homology (similarity) to consistently pair at meiosis. Therefore, few gametes will form.

Polyploidy can totally reverse this situation and restore fertility. As Benzing states, "In the tetraploid state, each chromosome in the egg or pollen mother cell has a homologue sharing the hybrid's diploid parents. These mother cells can now undergo normal meiosis and yield viable diploid gametes. The resulting tetraploid zygotes and embryos will perpetuate the polyploid event through potentially limitless future generations. . . . In time, the population may evolve a genotype that merits recognition as a bona fide species." (Benzing 1980, pgs. 197-198.)

Tillandsia Morphology and Natural Selection

According to taxonomists (Smith, 1934; Pittendrigh, (1948); Benzing, 1980), *Tillandsia* retains the least specialized floral structure in subfamily Tillandsioideae and, as such, could be considered the most primitive Tillandsioid genus. But if taxonomic assignment were based on vegetative characteristics, the extreme epiphytes of the genus might be considered the most highly advanced in the entire family, let alone Tillandsioideae.

Floral structure, which is used to assess relationships, has little bearing on moisture and nutrient procurement, drought endurance, or shade tolerance. Likewise, the vegetative parts of the plant, whether xeric or mesic, have little direct influence on reproductive structures. Each organ, controlled by different sets of genes, may be subjected to different selective pressures in the environment. In evolutionary terms, one can easily see why different body parts often evolve at different rates and why species with similar floral characteristics often have vegetative morphologies vastly different from species in the same subgenus.

The interest of most collectors has centered on the epiphytic, xeric species, but a majority of *Tillandsia* species are mesic, and many, such as *T. ferreyrae* L.B. Smith and *T. paniculata* Linnaeus, grow to be very large, often two or three meters in width and height. They are less widely available to the public and probably would not be popular if they were; the blooms are often nondescript. Sometimes mesic and/or high altitude species have exacting moisture, air circulation, and light requirements that make them difficult to cultivate, such as *T. biflora* Ruiz & Pavon and *T. multicaulis* Steudal. Others are less demanding and are correspondingly dispersed over a broader geographical range, as are most of the species presented in this book.

Changes in the climate of Latin America have forced numerous plant taxa to adapt rapidly to changing environmental conditions, specifically to extreme (although highly adaptive) epiphytism. How have they been able to do this? According to Benzing, *Tillandsia* precursors must have had the rudiments of an absorbing trichome or foliar hair amenable to refinement as a water- and nutrient-procurement device. The plants must also have been predisposed in other ways to the harsher existence of canopy life.

Benzing (1980) says, ". . .as aridity increased, the tank habit and thin, soft leaves became less adaptive while the serviceability of absorbing hairs, succulence and other atmospheric characters increased. With selective forces in the environment favoring the survival of the most xerophytic progeny, only time was required to promote the emergence of the air plants with all of their current adaptive refinements." Today's xeric species of *Tillandsia*, along with some orchid and cactus species, have become virtually the only plants to combine the indispensable vegetative and reproductive organs and systems necessary to epiphytically exploit harsh, oligotropic environments.

Let us now look at the functional significance of individual morphological features of plants in their never-ending adaptation to the opportunities presented by the forces of nature. There are many factors that make up the survival strategy for each population and much work remains to be done in unlocking the secrets so wondrously evolved over the ages.

Much of what follows is reasonable conjecture based on solid scientific principles and correlative studies with other plant families. Although interest is increasing, relatively little scientific work has been conducted on *Tillandsia*. Subjects such as population genetics, chromosomal and trichome correlations for the entire genus, floral evolution, and taxonomy are some of the areas open for continuing study.

Many factors have contributed significantly to the success of the more than 400 *Tillandsia* species. One of the most important in terms of genus diversity and radiation may be the ability of the plumose seeds to float on air currents. To a degree, distances *Tillandsia* seeds are able to travel are limited only by the strength of the air currents that carry them. Although most seeds do not travel far, some have probably traveled hundreds of miles. Seeds dispersed in this manner may found new populations.

Long-range dispersal inevitably brings the species under new selective pressures. On a large scale, most Latin American countries are a mosaic of diverse environments ranging from tropical jungle, to chaparral, montane desert, and forest. When seed-grown plants are able to survive in a new and different environment, they are subjected to an increase or decrease in the frequency or amount of moisture levels, light, temperature, wind, available pollinators, etc. The founder population must adapt through natural selection to accommodate these new conditions.

The ability of the atmospherics to grow without soil in harsh conditions has opened up vast new habitats where the plants grow unchallenged in the competition for space. Benzing (1980) has stated that the most significant adaptive shift in *Tillandsia* is toward extreme epiphytism. He also found the "movement from equable to harsh habitats was accompanied by gradual, but ultimately elaborate, modifications of both the form and function of the basic Tillandsioid body." It is the combination of these factors and more that has allowed the atmospherics to so successfully occupy xeric habitats.

Among these modifications are: development of a thick adaxial hypodermis (tissue between epidermis and mesophyll) in which to store water and salts, increasing number and shield complexity of the absorbing trichomes, decrease in leaf number, decrease in root development, decrease in overall plant size and impoundment capacity, reduction in the number of flowers and fruits per shoot, and, according to Gilmartin and Brown (1985), an increasing tendency for CLEISTOGAMY (self-fertilization; kly-STA-ga-mee). In addition, the xeric species have increased the efficiency of water and mineral use. Gardner (1982) states, "The highly evolved absorptive trichomes contribute to efficient water and mineral procurement, while crassulacean acid metabolism (CAM), characterized by nocturnal stomatal opening, enhances water economy. These characters combine to allow the species to colonize habitats that would be unsuitable to most epiphytes."

The decrease in leaf and root number along with overall plant size allows the shoot to channel a greater proportion of its energy into seed production. One of the key factors in the *Tillandsia* success story has been the large number of seeds produced for a small plant with such limited resources for reproduction.

Xeric tillandsias have evolved to be stress tolerant because extreme epiphytism is possible only under certain, very rigid criteria. In contrast, atmospherics are poorly designed by their succulence, CAM method of photosynthesis, and other factors to grow large quickly.

Tillandsias have entered previously unoccupied habitats and prospered. Other types of plants have not followed. Why? According to Benzing, ". . . the major problems of extreme epiphytic life do not revolve around defying gravity or drought *per se*, or sustaining adequate dispersability, but rather of mounting a properly scheduled reproductive effort of considerable magnitude against a background of overall low productivity.." This summarizes what has been previously stated; namely, that it is a combination of all these genetic and environmental factors that has allowed xeric tillandsias to accomplish what they have—large numbers of plants of varied species, widely dispersed, that are often very distinct in appearance from one another. With the exception of the foliar trichome, that in the xeric species is refined far beyond those of any other type of plant, all other vegetative strategies incorporated by xeric tillandsias are found in other families.

More and more attention is now being given to understanding the role of natural selection and observing the myriad ways xeric tillandsias have evolved adaptations, such as epiphytism, size, shape, trichomes, and CAM metabolism. Tillandsias are widely dispersed over the Americas, but according to Stebbins (1966), geographical distances separating species do not, by themselves, account for variation. Instead, differences among species, as we have seen, are the result of evolutionary responses to environmental pressures.

On a small scale, the environment is a collection of microhabitats, each of which presents its own set of ecological conditions. At one end of the spectrum, even a tree provides differing levels of shade, moisture, etc. Some species, such as *Tillandsia ionantha* v. *ionantha* and *T. brachycaulos*, flourish on the same tree branches. And while there is great population overlap, *T. ionantha* tends to grow in the more exposed positions of the tree, such as at branch tips, while T. brachycaulos is encountered in more shaded parts of the tree. In a larger context, these microhabitats can be grouped into ecosystems that are subject to many of the same environmental pressures, such as day length, relative temperature, and moisture levels.

The overall environmental system in Latin America has not been a model of climatological consistency as had been thought. Instead, the climate of much of the area has frequently oscillated between humid and arid phases. This has had a direct bearing on the evolutionary development of *Tillandsia*

Taxonomically, *Tillandsia* subgenera are separated from each other by differences in floral structure. With regard to the seven currently recognized subgenera, some of them possess both strongly xeric and mesic forms. Why is this so? The difference in morphology between mesic and xeric species is so great, it would seem reasonable that mesic forest-dwelling species would have evolved into the same or similar subgenera on the one hand, while the xeric taxa would have evolved to be grouped together on the other.

Directional Selection

An answer suggested by Gilmartin (1983) lies in the climatological history of South America. Scientists have recently learned that the continent has undergone a number of humid and arid phases, none of which has lasted for longer than 15,000 years. Livingston (1980) says, "Where the rich biota of Amazonia was once attributed to a constant physical environment of great antiquity, it is now attributed to a pulsing physical environment that periodically restricts the forest biota to limited regions where genetic drift and rigorous selection of small populations rapidly generate new taxa." Much evolutionary radiation can be attributed to this example of two alternating types of DIRECTIONAL SELECTION. Directional selection occurs when the environment is changing progressively in a particular direction—that is, drier/wetter, hotter/colder. It produces a regular change of the population in one direction with respect to certain adaptive characteristics. In other words, species adapt biological systems (morphological and physiological change) to the widespread xeric conditions of an arid environment. When xeric conditions start to be interrupted by an expanding mesic climate, the isolation fixes new species.

Gilmartin elaborates, "During an arid phase, the xeric forms of each subgenus are favored for migration and establishment, while the mesic forms are isolated (in small populations) from one another in refugia. Isolation favors the evolution of new species as herewith the isolated mesic forms during the dry phases. On the other hand, the opposite occurred during the humid phases. The xeric species became isolated in dry refugia and further speciated while the mesic species dispersed and became established in a wider range until the next stage in the cycle of climatic changes."

As an illustration, assume that during a humid period, a species inhabited a large uninterrupted area of forest covering two mountains and the valley between them. As the climate gradually changed to become more arid, the valley might have lost its mesic quality first, thereby creating an isolated *Tillandsia* population of the same species on each mountain. As the climatic change continued, and the mesic habitats shrank, the two populations would be further divided into smaller isolated groups that would be susceptible to speciation. In contrast, the xeric species would spread into new habitats during an arid period, and small, wind-dispersed founder populations could also become established, thereby promoting speciation.

Gilmartin continues, "At the generic and subgeneric levels, the evolutionary line evolved into both xeric forms and mesic forms. . . . The evolutionary trend within the genus *Tillandsia* must have been *from* the mesic Allardtia-like ancestors *to* the (evolutionary) branches, giving rise only to xeric forms. This kind of reasoning would place the xeric forms at branch tips (of the evolutionary tree). . . ."

In summary, genetic speciation and/or adaptive change occurs through the processes of natural selection. Natural selection operates as tillandsias adapt to new environmental pressures. This is possible because sexual mating, especially in isolated populations, provides new combinations of genetic traits from a large and changing gene pool. The gene pool changes because of genetic mutation, changes in chromosome structure and number, and genetic recombination. Pleiotropy is another method by which change can occur.

Hybridization, back crossing, and genetic stabilization of new hybrids, known as introgression, is another method through which speciation and evolution can occur rapidly occur. Although polyploidy does not seem to be much of a factor in Bromeliaceae, it has figured strongly in the evolution of 40-50% of angiosperm species.

These genotypic changes are able to phenotypically manifest themselves as new species in small, reproductively isolated populations. In *Tillandsia*, current theory holds that these isolated populations have been created largely as a result of the expansion and contraction of arid and humid environments in Latin America, along with certain adaptive traits such as the wind-dispersal capability of the seed. Because of the great complexity of both the plants and their environment, the interaction between them can alter populations in a multitude of ways.

Literature Cited:

Anderson, E. 1949. *Introgressive Hybridization*. Biological Review; 28: 280-307.

Benzing, D.H. 1980. *The Biology of Bromeliads*. Eureka, Calif.: Mad River Press.

Bromeliad Society, Inc. 1977. *International checklist of Bromeliad Hybrids*. Los Angeles: The Brom. Soc. Inc.

Darwin, C. 1859. *On the Origin of Species*. Norwalk, Connecticut. Easton Press, pgs. 45-46.

Gardner, C.S. 1982. *A Systematic Study of Tillandsia Subgenus Tillandsia*. College Station. Ph.D. dissertation. Texas A & M University.

__________. 1984. Natural Hybridization in *Tillandsia* subgenus *Tillandsia*. Selbyana 7:380-393.

Gilmartin, A. J. 1972. *The Bromeliaceae of Ecuador*. Germany: J. Cramer.

__________. 1983. Evolution of mesic and xeric habits in *Tillandsia* and *Vriesea*. Systematic Botany Vol. 8, No. 3, pgs. 233-242.

Livingston, D.A. 1980. History of the tropical rain forest. Paleobiology 6:243-244.

Mayr, E. 1963. *Populations, Species, and Evolution*. Cambridge, Massachusetts: Belknap Press, Harvard University.

Pittendrigh, C.S. 1948. The Bromeliad-Anopheles-Malaria complex in Trinidad. The Bromeliad flora. Evolution 2:58-89.

Smith. L.B. 1934. Geographical evidence on the lines of evolution in the Bromeliaceae. Bot. Jahrb. 66:446-468.

__________. and Downs, R.J. 1974. *Flora Neotropica. Monograph No. 14. part 1. Pitcairnioideae*. New York, N.Y. Hafner Press.

__________ and __________. 1977. *Flora Neotropica. Monograph No. 14. part 2. Tillandsioideae*. New York, N.Y. Hafner Press.

Stebbins, G.L. 1966. *Processes of Organic Evolution*. Englewood Cliffs, N.J. Prentice-Hall, Inc.

__________. 1971. *Chromosomal Evolution in Higher Plants*. Menlo Park, California. Haddison-Wesley Publishing Co.

Wright, S. 1982. Character change, speciation, and the higher taxa. Evolution 36:427-443.

__________ . 1985. Evolution in Mendelian populations. Genetics 16:97-159.

Water and Food Transport

With an understanding of how a *Tillandsia* moves water and foodstuffs through its constituent parts, the reader will better comprehend the significance of various outward changes in a plant's appearance. This will help the grower to monitor the plant's moisture status, which is important for the health and development of specimen plants.

Most of the food transported about the plant exists as CARBOHYDRATES, such as the sugar SUCROSE, which cells break down in order to get needed energy. In addition, AMINO ACIDS and many other organic substances are transported. Carbohydrates are produced during the photosynthetic process in green-colored cells. Green cells produce more carbohydrates than they need for their own use. All living cells use carbohydrates produced by photosynthesis to obtain energy for maintenance and growth. Energy SINKS, such as shoot tips, root tips, flowers, developing fruits, new offsets, and immature leaves, do not have green cells, so they must import photosynthate from leaves that can use light to produce this material.

The leaves are called SOURCES because they manufacture and export energy in the form of carbohydrates to the sinks. The relationship of sinks and sources is constantly changing as the plant goes through its life cycle, and new organs are formed as older ones die. At first, a young leaf is an energy sink because it uses more energy than its green cells produce. When about half size, the leaf begins producing more energy than is being used for growth, thus becoming an energy source.

Water and food move through the different parts of a *Tillandsia* by way of vascular tissues, as in other land plants. The water-conducting tissue, termed XYLEM (zye-lem), comprises VASCULAR BUNDLES of cells that are called TRACHEIDS (TRAY-key-ids) and VESSEL ELEMENTS. Foodstuffs, mostly in the form of sucrose, are transported in tissue called PHLOEM (floam). The cells that comprise the vascular system for foodstuff transport are called SIEVE-TUBE ELEMENTS.

Tracheids, vessel elements, and sieve tube elements are elongated and positioned end to end in long series. The walls of these structures, as well as those of all plant cells, are partially constructed of microscopic fibrils (small fibers) consisting of CELLULOSE molecules. Cellulose is, in turn, made of many smaller glucose molecules laid end to end.

Space between the fibrils often has other substances that give a particular group of cells help with their function. Those cells responsible for major structural support, such as the walls of xylem and sometimes phloem, have a material called LIGNIN in the interstitial spaces of the fibrils. Like cellulose, lignin is a large molecule in the polymer class.

Vessel elements and tracheids of xylem tissue die once they mature, and thereafter fulfill their role as water conduits. If alive, the cell centers would be filled with the living protoplast, which would inhibit water movement. With the cells dead, the protoplast is gone and only the hollow SCLERIFIED (a cell wall heavily impregnated with lignin) cell walls remain.

The sclerified xylem cells, which make up the bulk water conduit system, are of two basic types. Tracheids are more primitive than vessel elements in structure and efficiency and less adequately designed for efficient water transport—tracheids are longer and narrower. In addition, tracheids are not hollow at the ends but closed. They have thin spots called PITS along the cell walls and at each end. Water moves from tracheid to tracheid across these aligned pits. The groups of pits allow for adequate rates of water to flow among them, as long as the plant is watered sufficiently.

Primitive vessel elements are like tracheids, but in addition to pits along the lateral cell walls, there are small openings in the PERFORATION PLATES (flat ends of the cells). As vessels become more advanced, the transverse ends that abut each other become shorter and wider with progressively larger openings. In the most advanced vessel elements, perforation plates are dominated by a single, large pore. These pores allow water to flow at higher rates.

Conifers, the majority of ferns, and some primitive flowering plants rely exclusively on tracheids for water transport. Most bromeliads have a mixture of tracheid and primitive vessel elements. Surprisingly, plants in *Pitcairnia*, a genus in the most primitive bromeliad subfamily, show the most advanced xylem tissue in the entire family.

The atmospheric tillandsias have a higher concentration of tracheids and primitive vessel elements. One might have thought that these plants would have the most sophisticated arrangement for water transport because of the environmental stress routinely placed upon them. This stress would select for plants better adapted to move water. However, according to Benzing (1980), similarly primitive xylem cells are known to work well in other groups of plants that have to endure drought cycle climates. He feels that the broad, short vessel element of the more advanced type, which is so well designed for large water flow, is more prone to CAVITATION (formation of gas bubbles), thus impeding water flow in plants under high water stress.

In addition, because water is passed directly from the epidermis to the hypodermis (water storage tissue) via the trichomes, it does not have to travel far internally, as it would if the water were absorbed through the roots. An advanced internal transport system is not necessary. Xeric species of *Tillandsia* have reduced areas of vascular tissue and large areas of hypodermis relative to other xeric terrestrials. This aids greatly in droughty periods. Epiphytic xerophytes of the families *Orchidaceae*

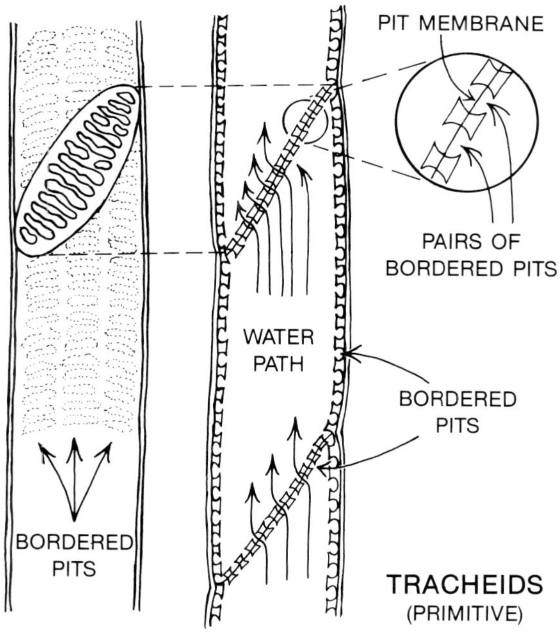

Fig. 6.4

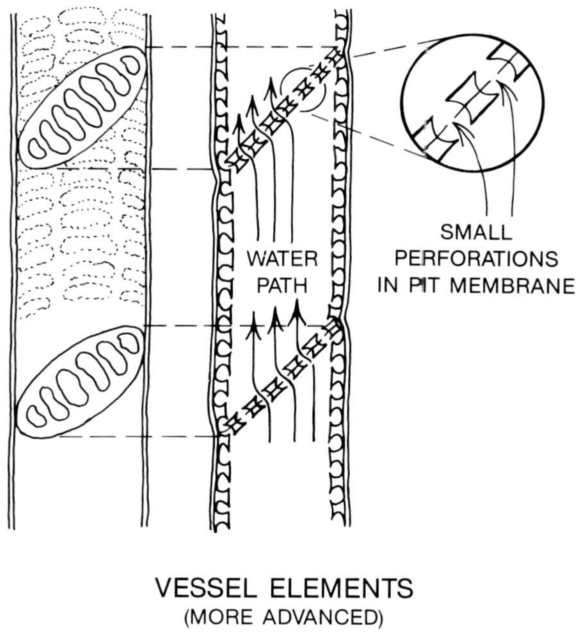

Fig. 6.5

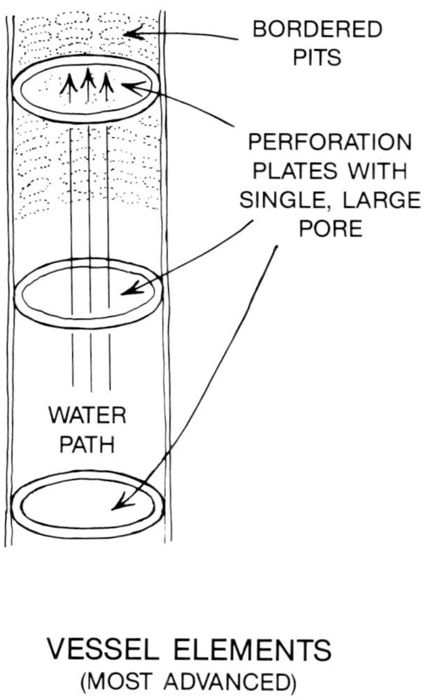

Fig. 6.6

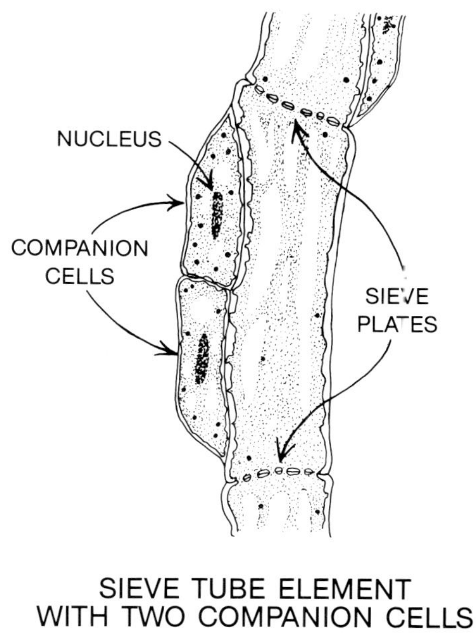

Fig. 6.7

(redrawn from Benzing)

and *Cactaceae* also share these methods of water absorption and increased hypodermal tissue.

SIEVE PLATES at each end of the sieve-tube element are perforated with small openings. These orifices allow concentrated solutions (mostly sucrose) to pass from one element to the next. Usually, the organic substances are moved from the sources to the sinks. Less is known about the structure and function of the sieve-tube element than any other type of plant cell.

Each sieve-tube element has one or more narrow parenchyma cells called COMPANION CELLS, closely associated with mutual cytoplasmic connections. The companion cells are alive and have protoplasts containing abundant mitochondria, a nucleus, and other ORGANELLES (subcellular bodies). Phloem movement is quite slow compared to the water flow in xylem tissue.

Cellular Water Transport

As with all living creatures on earth, the entire plant must function as an integrated unit from the submolecular level to the biological coordination of each leaf, root, and flower.

In the section on Water and Food Transport, we saw that water and nutrients are moved in bulk through a plant via vascular bundles. Water, nutrients, and stored energy in the form of carbohydrates must also be able to penetrate and move through living cells requiring supply. How do the individual cells of a plant, or even organelles within cells get the water and nutrients they need to function properly? How is a leaf or plant able to fully recover from severe dehydration? Conversely, at what point does a leaf or plant receive too much water? What happens if it does, and by what signs does one know that it has been watered too much?

Water is absorbed into atmospheric tillandsias chiefly through specialized trichomes. These are described in detail in the section on Foliar Trichomes.

Water is the universal biological SOLVENT, and life as we know it could not exist without it (a solvent is the medium in which a substance, or SOLUTE, is dissolved). Water is able to support life because so many essential minerals and nutrients will dissolve in it without altering their innate structure. A healthy *Tillandsia* consists of 70-90% water.

A plant cell is a living unit composed of numerous parts. The somewhat rigid cell wall is freely permeable to water and various minerals, nutrients, and other solutes. Inside the cell wall there is a PLASMA MEMBRANE that is DIFFERENTIALLY PERMEABLE.

A differentially permeable membrane allows small molecules such as water to pass through it, but is not permeable to some IONS or larger molecules such as PROTEINS. It is also a membrane through which substances pass at different rates. The movement of water across a differentially permeable membrane is called OSMOSIS. There is potential for impure water inside a cell to draw pure water into itself in order to balance the level of solute concentration on the two sides of the membrane. This impure water contains potassium sugars, organic acids, and other solutes.

To the inside of the plasma membrane is a watery matrix called the CYTOPLASM. The cytoplasm contains all the cell organelles. Most of the cellular activity goes on in the cytoplasm because of the various functions and responsibilities of the organelles.

To the inside of the cytoplasm is another differentially permeable membrane called the TONOPLAST, which bounds the VACUOLE. The tonoplast controls the flow of water and substances into the vacuole. The vacuole is a water-filled vesicle occupying much of the PROTOPLAST (a living cell minus its wall).

The pressure to draw water into a cell, which increases cellular volume, is called OSMOTIC POTENTIAL (usually represented by the Greek letter *pi*, π). Obviously, water cannot be drawn into a cell indefinitely. TURGOR PRESSURE, P, arises from the positive mechanical pressure generated by the resistance of the cell wall to expansion as the inflow of water seeks to increase cellular volume. Normally, enough P is created to equalize a cell's water potential (represented by the Greek letter *psi*, ψ) so that the resulting pressure balance approaches zero. This balance maintains the rigidity of the cell wall.

A cell can partially dehydrate without permanent injury. How much it can desiccate and fully recover is not easily measured. As can be seen in a fully turgid cell (Fig. 6.8 A), the protoplast is pushing against the cell wall. In the next drawing, the cell has lost so much water that its turgor pressure is nearly zero. At this point the cell is only about 90% turgid. It has low water potential, because there is almost no P pushing out to offset the osmotic potential, π, which means the cell is drying out. In this example, the cell would take up water if it were available in the interstitial spaces among the cells.

The third drawing shows a cell that is severely dehydrated. The protoplast has been drawn away from the fully relaxed cell wall. When this happens the cell usually will not recover, no matter how much water is made available to it. This is because the cell has suffered mechanical damage. Fortunately, PLASMOLYSIS (severe dehydration) can be avoided because it can be visibly seen by the attentive grower, as will be discussed in the next section.

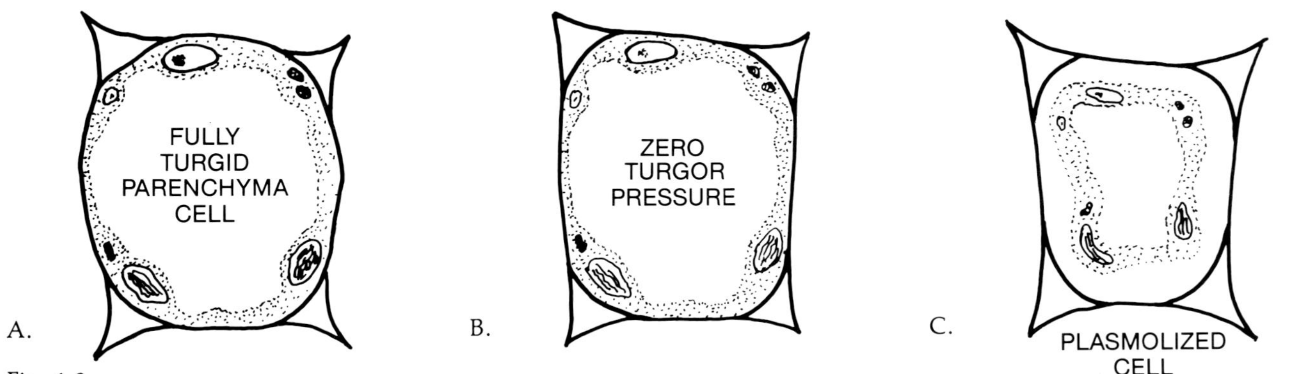

Fig. 6.8

(redrawn from Benzing)

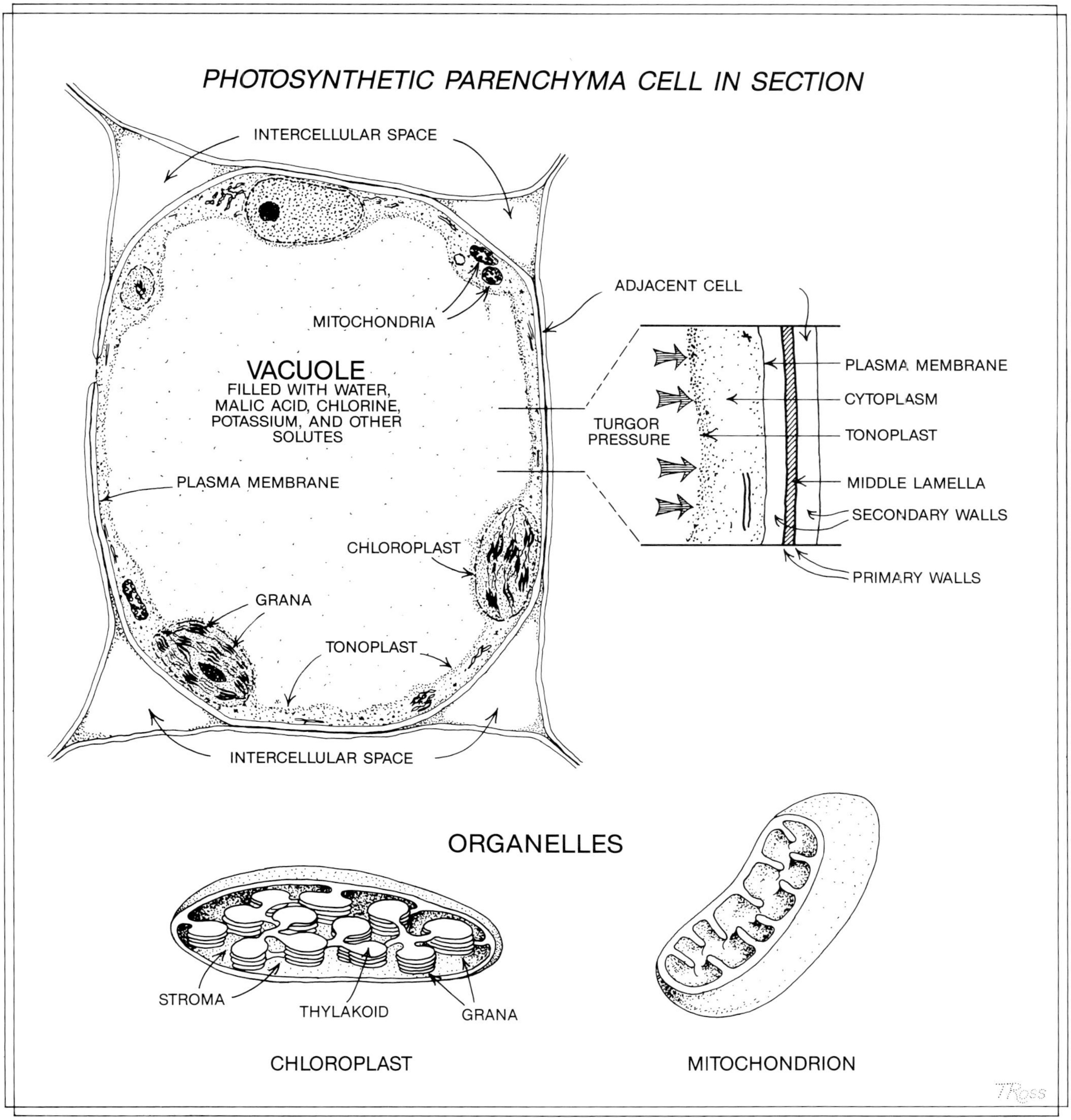

Fig. 6.9 (redrawn from Benzing)

Water and Moisture Aspects in Growing Tillandsias

One of the most significant adaptations of xeric tillandsias (which includes the majority of species presented in this book) is their ability to survive without water for long periods of time—even months in extreme cases. The way in which atmospherics survive through these dry periods is remarkable.

Visual indications of desiccation are severely involuted leaves in species such as *Tillandsia aeranthos* and *T. brachycaulos*, or the twisting of leaves into tight ringlets in species such as *T. duratii, T. caput-medusae*, or *T. streptophylla*. With a little experience, close inspection will quickly reveal whether or not a plant is beginning to dehydrate. A soaking for a day or two in tepid water will rehydrate the specimen.

For any plant, the ability to retain water is influenced directly by the temperature and humidity of the surrounding air. In most plants, water lost from TRANS-

PIRATION is quickly replenished by water absorbed through the roots and brought up to the leaves via the xylem vascular tissue. But the method of xeric tillandsias is different. Not only do they absorb water through their leaves instead of their roots, but many of them also undergo periods of drought. As a consequence, the plants must be conservative about water use.

Understanding a particular environment is essential for growers who want their plants to flourish. Light, water, temperature, and fertilizer are the major factors that most affect the health of the plants. These can be important assets in growing magnificent plants.

Within a broad range of atmospheric conditions in the natural environment, a *Tillandsia* has the ability to control transpiration—to a point. In a severe drought many tillandsias may perish because too many cells experience plasmolysis. I have seen examples of this on a large scale in Peru with populations of *Tillandsia paleacea* and in Argentina with *T. xiphioides*.

If a desiccated plant does not recover after a one- or two-day submersion in water, one should chalk up the loss to experience. Virtually every grower around has experienced the pain of losing a prized specimen.

Once in a while someone will glumly proclaim, "Oh, my plant was doing fine until I watered it. Then it fell apart." As the years go by, experience has shown that the atmospherics LIKE WATER. They will survive nobly without it for long periods of time—especially in humid environments—but why torture them to the limits? Grown outdoors with plenty of light, it is hard to overwater most tillandsias in a warm, low-humidity environment. My strong hunch is that the plant in question did not die from overwatering. Because it was almost never watered, it had already died from the effects of plasmolysis. When the plant was finally watered, the dead plant fell apart.

Some species with dense leaves, such as *Tillandsia argentina* and

Fig. 6.10 A severely dehydrated *Tillandsia aeranthos*.

Fig. 6.11 Same plant after a 14 hour submersion.

Fig. 6.12 Desiccated *Tillandsia streptophylla*.

Fig. 6.13 *Tillandsia tectorum* that has been watered too frequently and/or not received sufficient light.

T. ionantha, often grow in clumps. Care must be taken with these species to permit the wet leaf epidermi to dry between waterings. When wet, stomatal gas exchange and photosynthesis are inhibited. The liquid film of water held on the epidermal surface by the trichome caps acts as a barrier to this vital respiratory process. For a period of time, probably a day or two, this is not critical. But a point is reached when the plant is starved for oxygen, and just like a person, it will suffocate.

For those in humid climates, watering is a different situation altogether. With hot, stifling humidity, dehydration is usually not the major problem. Instead, plants can easily be kept wet for too long. The metabolic rate is faster with higher temperatures, which means that more oxygen is consumed by the plant. This in turn means that the period of time the leaves can survive without respiring, such as when the epidermi are wet, is reduced.

Stagnant air is another contributor to the demise of many atmospherics. Fresh air greatly shortens the period of time needed for an epidermis to dry. Warm, humid, stagnant air also produces an environment conducive to the proliferation of bacteria, fungi, and mold.

Fig. 6.14 Closeup shows damaged epidermal scales. They should be dense with long wings perpendicular to the leaf surface.

Rot from suffocation is easy to detect—but only when it has progressed too far and at least some leaves have become brown. Many people have set a plant down on a bench with the intention of mounting it "soon," only to discover after a few weeks that the plant has been damaged or even had died.

If this occurs, and it is noticed soon enough, the upper portion of the plant may survive if it looks healthy. All dead material should be immediately removed. Besides dead leaves, any brown tissue in the stem must also be excised. A sharp knife can be used for this purpose. The affected portion may be cut away in the same manner one removes a bruise from an apple. The plant can then be put aside to dry thoroughly. If done soon enough, it will recover or at least produce a number of offsets.

Watering tillandsias indoors presents its own special set of circumstances and is discussed in the section on Indoor Care. Watering plants indoors assumes great importance because of the generally unfavorable light, humidity, and air movement conditions.

Photosynthesis

From the simplest bacteria to *Homo sapiens*, all living organisms must possess the ability to convert food into energy. Those that utilize inorganic substances to manufacture food, e.g. light, water (H_2O), carbon dioxide (CO_2), and mineral elements, are called AUTOTROPHS. Tillandsias and most other plants are autotrophs. Animals, including both herbivores and carnivores, fungi, and nongreen parasitic plants, such as some mistletoe, are HETEROTROPHS. They must assimilate other organic substances—either plants or animals—and as such are classified as either PREDATORS, PARASITES, or SAPHROPHYTES (consumers of dead organic matter).

PHOTOSYNTHESIS is a critical process by which green plants harness and utilize radiant energy from the sun to manufacture food. This is extremely important

Fig. 6.15 *Tillandsia tricolor* with numerous offshoots, the result of center rot to the parent plant.

Fig. 6.16 *Tillandsia caput-medusae* with some leaves dead from over exposure to water.

because it is the basis for the entire food chain. Without the ability of plants to photosynthesize, life would cease to exist on earth.

Although much remains for science to uncover about the complex procedures that take place during photosynthesis, basic details are known. How the atmospherics manage to utilize the sun's energy in what we would normally consider to be hostile, dry environments (especially for epiphytes) is the subject of this section.

The portion of light that humans can see is called the VISIBLE SPECTRUM. It closely corresponds to the portion of solar radiation used by plants to manufacture carbohydrates and is called PHOTOSYNTHETICALLY ACTIVE RADIATION (PAR).

There are three major photosynthetic mechanisms by which higher plants create food using PAR, water, and atmospheric carbon dioxide, but tillandsias use only two of them. Soft-leaved, mesic tillandsias use one method, and the harder, stiffer-leaved atmospherics another. The manner in which mesic species convert PAR into food will be discussed first.

C3 Photosynthesis

Species such as *Tillandsia yunkeri* L.B. Smith, *T. leiboldiana* Schlechtendal, and *T. punctulata* with a biosynthetic process called the CALVIN-BENSON PATHWAY, which is named after the two scientists who worked out the chemistry of the reactions involved. Tillandsias that use only this pathway are called C3 PLANTS. This is the method usd by most other green plants on earth to manufacture food.

For all plants, photosynthesis takes place only in cells that have organelles called CHLOROPLASTS. These photosynthetic cells are always found near the plant surfaces. Chloroplasts capture PAR, which is then absorbed by CHLOROPHYLL, a pigment found in chloroplasts. Chlorophyll is what makes leaves green.

Chlorophyll is contained in flat, multilayered membrane systems called THYLAKOIDS (THIGH-la-koids).* Some thylakoids are in turn stacked up in piles. The liquid in which the thylakoids are embedded is the GRANA or STROMA. Stromal enzymes utilize energized molecules, such as ATP (adenosine triphosphate), to catalyze atmospheric CO_2 into organic molecules and simple carbohydrate sugars such as glucose.

Part of this stored energy is transferred to organelles called MITOCHONDRIA (my-toe-CON-dree-a). There may be hundreds in a single cell. Mitochondria produce energy by breaking down the stored sugars into smaller chemical byproducts, namely, H_2O and CO_2. The energy released in this process is exported to various other organelles to perform work.

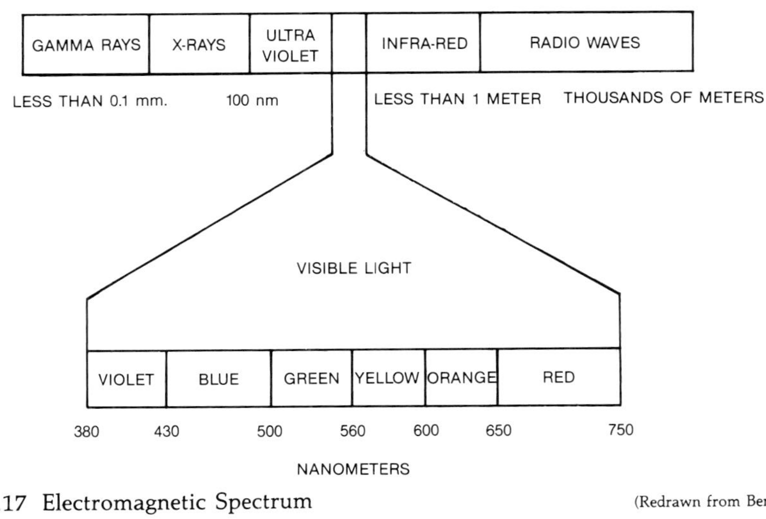

Fig. 6.17 Electromagnetic Spectrum (Redrawn from Benzing)

This breakdown of sugar molecules into simpler constituents in an extremely complex series of chemical reactions is called RESPIRATION.

Some of the stored energy cannot be kept in the mitochondria, so it is kept in the form of starch grains and oil droplets in the cytoplasm of the cell. These starch grains are eventually converted to sugar, while the fats and oils are broken down into small fragments and fed midway into the respiration process.

Having discussed PAR and how it is used by the plants, it is now time to consider the elemental building block of life, CARBON. Carbon atoms combine with other elements to form the basic molecules that make up all life on earth. CARBOHYDRATE means composed of carbon and the elements of water. Carbon is taken into the abaxial leaf and bract pores called STOMATA (singular, STOMA). It is brought in as part of the atmospheric molecule, CO_2.

The term LIGHT REACTION is used to describe the capture of light energy and the stabilization of that energy into energized molecules (ATP). During the light reaction process, oxygen (O_2) is liberated as a byproduct. It eventually diffuses out of the leaf into the surrounding atmosphere.

Throughout the daylight hours, in a DARK REACTION process, CO_2 in the leaf is consumed by the Calvin-Benson pathway and fed directly into the multistep synthetic route to make carbohydrates. It is called DARK REACTION because technically, the Calvin-Benson cycle does not require light directly.

O_2 is consumed and CO_2 is generated as a byproduct of cellular respiration within the mitochondria. At this time, carbohydrates and other organic compounds are broken down and used to meet the energy requirements for cell maintenance and growth. This use of energy reserves occurs twenty-four hours a day, but during the daylight hours photosynthetic activity refixes the byproduct CO_2 through the Calvin-

**See* Fig. 6.9, pg. 187.

Benson pathway before it can escape from the leaf.

To summarize, C3 plants consume CO_2 and evolve O_2 during the day, and take up O_2 and liberate CO_2 at night. This system for respiring and converting PAR into useful foodstuffs is efficient, but it would prove fatal to any plant without access to abundant supplies of water. Even so, it works well for a majority of the earth's plants.

C3 plants require large amounts of water, because during the day the plants are photosynthesizing and the stomata are open. Although the stomata may represent only 1% of the epidermis, sufficient water can evaporate out of them to cause serious injury to a plant if it does not replenish the losses. This is especially true in environments that are either hot and/or dry. These plants must have access to continuous water supply via root absorption; or, as in the case with tank-type bromeliads, through foliar uptake from impoundments. Imagine what would happen to many vegetables or houseplants such as *Coleus* or *Begonia* if they did not receive water for two weeks in the hot summer!

CAM Photosynthesis

What happens to a xeric *Tillandsia* that lacks a root system or impoundment tank for the uptake of water? The plant grows on a desert cactus, the sunlight is intense, it receives rain only a few weeks of the year, and there is no friendly caretaker to water it with a hose.

Scientific experiment (Benzing and Dahle 1971) has shown surprisingly, that instead of taking in CO_2 by day, xeric *Tillandsia* species absorb almost none. Equally astonishing is the fact that while C3 plants discharge CO_2 at night, xeric tillandsias take in large quantities of it. How can this be?

Understanding CAM photosynthesis requires more specific knowledge of how the epidermis behaves with relation to the exchange, or diffusion, of water and gases between the leaf and air.

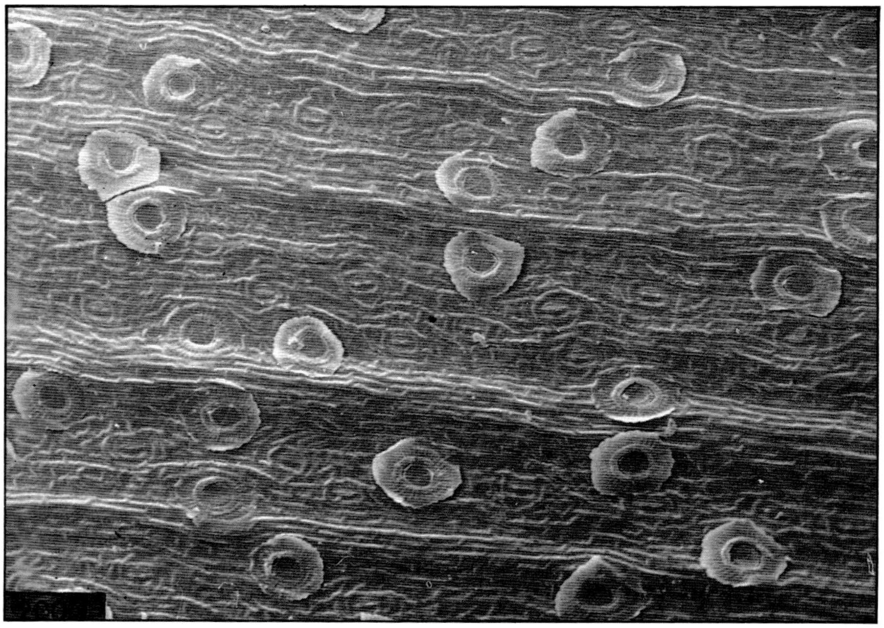

Fig. 6.18 Abaxial epidermis of *Tillandsia complanata* Bentham. Note closed stomata and more visible trichomes (X100; photo Gross).

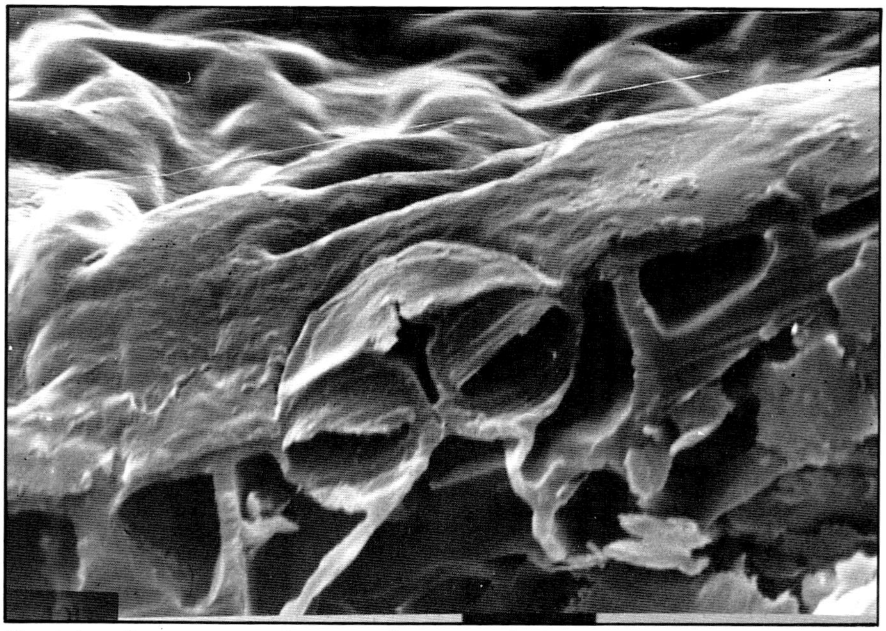

Fig. 6.19 Cross section of a stoma from same leaf showing closeup of guard cells (X2,000; photo Gross).

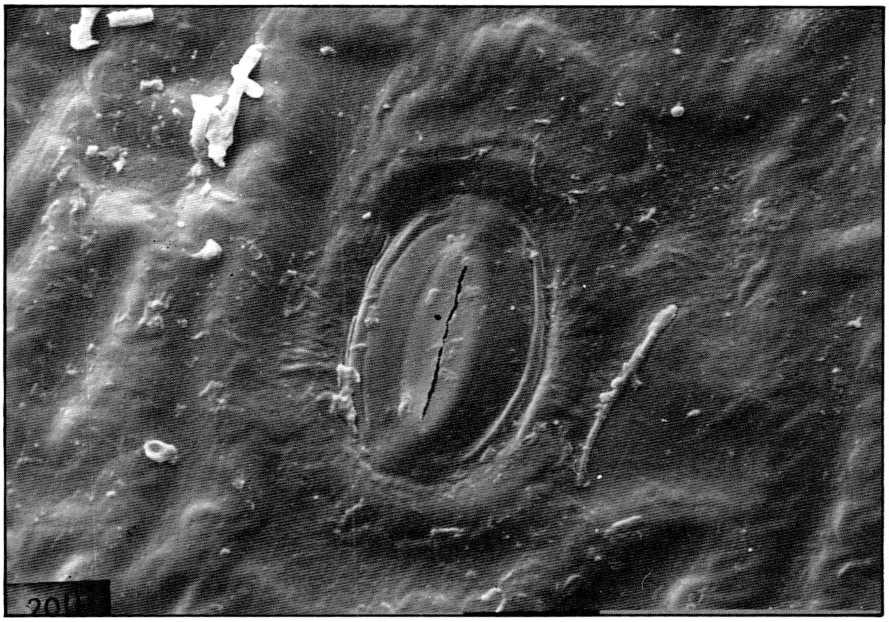

Fig. 6.20 Closed stoma from leaf of *Tillandsia complanata* (X1,000; photo Gross).

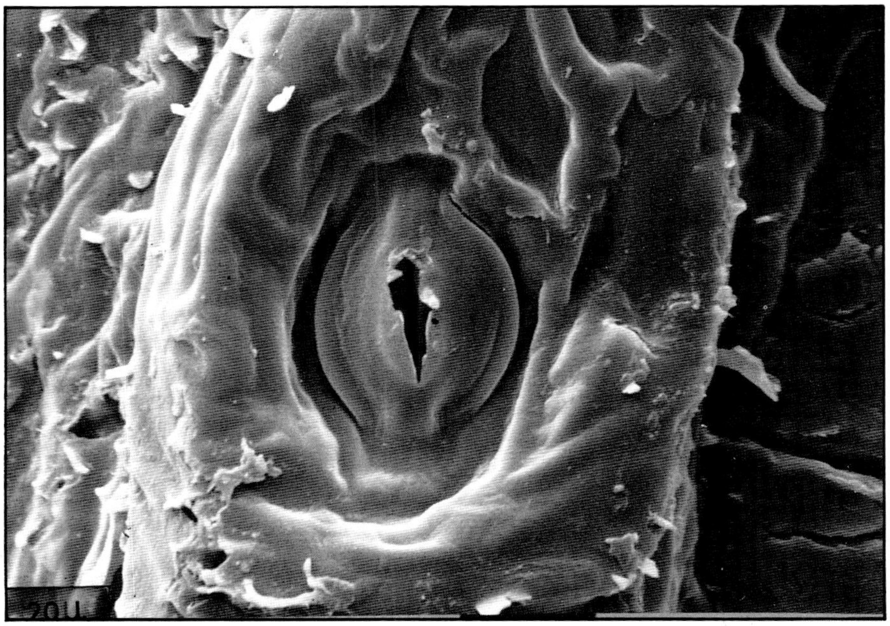

Fig. 6.21 Open stoma from leaf of *Tillandsia stricta* (X1,000; photo Gross).

DIFFUSION is the tendency of one substance to mix with another. It happens because of the kinetic properties of matter; molecules of all substances—gas, liquid, or solid, vibrate constantly at temperature dependent rates. The molecules of a gas move much greater distances at a given temperature than do those of a liquid or solid and those of a liquid more rapidly than those of a solid.

Plant scientists are interested in two diffusion processes. One occurs when a gas diffuses through another, such as when molecules of water evaporate out of a leaf into the surrounding atmosphere, which is transpiration. The other involves the diffusion of dissolved solids through aqueous solutions as discussed in the section on Water and Food Transport.

As previously stated, leaves have pores called stomata that allow the greater part of gas exchange to take place. Each stoma consists of a circular cluster of cells dominated in the center by two aligned, bean-shaped GUARD CELLS which are firmly anchored at both ends. The long inside walls of these cells can bow out laterally with an increase in turgor pressure. This bowing creates the stomatal opening.

When the guard cells of the stomata are closed, gas exchange is greatly reduced because, like a manhole cover in place, there are no open gaps and the surface is tightly sealed. A small amount of gas exchange and transpiration does occur because of the inherent permeability of the epidermal surface.

When the stomata are open, gases move freely in and out of the leaf. Remember from the section on Water Transport that atmospheric conditions determine the amount of gas exchange and also the rate of water loss. With higher temperatures and/or lower humidity, there is an increased rate of transpiration.

Tillandsias control the opening and closing of their stomata, thereby regulating the amount of epidermal porosity. The greater the porosity, the greater the potential for gas exchange, but, as a tradeoff, there also will be more transpiration (water loss). This is of critical importance, as we shall see shortly. The development of a mechanism, that is the highly advanced trichome caps, to deal with the rapid rate of transpiration in xeric environments is one of the evolutionary adaptations of the atmospherics.

The mechanism by which bromeliads in general, and tillandsias in particular, control leaf surface porosity is explained by Benzing (1980). Suffice it to say here that the stomata are sometimes open, sometimes closed, and that guard cell movements are actively controlled by the organism.

How does the plant photosynthesize if the stomata of atmospherics are closed during the day so the leaves cannot absorb CO_2? It does so by opening the stomata at night. The open stomata allow $C0_2$ to enter and diffuse into the leaves where it is absorbed by cytoplasmic enzymes. Sufficient PAR for the photosynthetic process was captured during the previous daylight hours. CO_2 absorbed and fixed at night is synthesized into MALIC ACID, a common fruit constituent. The malic acid is then moved to the water-filled vacuoles of the cells, where it is stored.

Soon after sunrise, with the stomata now flaccid and the leaf surfaces effectively sealed, the stored pools of malic acid are partially degraded by enzymes. This process liberates CO_2 into the leaf interior, where it remains trapped by the nearly impervious epidermis. As sunlight reaches the leaves, photosynthesis occurs, fed by the internally regenerated CO_2.

The rest of the process is the same as for C3 plants. By day's end, the stored malic acid has been used up. But with the day's photosynthetic production there is enough energy to repeat the procedure that night, along with sufficient energy to sustain growth and maintain tissue. The plant produces enough energy to grow, bloom, and reproduce by this method. Tillandsias are not unique in this respect. Succulent plants including cacti and other desert plants operate in similar fashion.

With this metabolic process, known as CRASSULACEAN ACID METABOLISM (cras-sul-AY-see-an) or CAM, xeric tillandsias consume and evolve $C0_2$ and 0_2 while C3 mesic bromeliads are doing just the opposite. What advantage is achieved by operating this way? How does it help them overcome water loss in harsh environments?

Xeric *Tillandsia* species are able to keep their stomata closed during the day when the sun's heat would cause extensive transpiration. This ability to "close down the hatches" during the daytime is an evolutionary leap forward and essential for plants living in environments with long rainless periods. These CAM plants lose water much more slowly because transpiration occurs at night when it is cooler and usually more humid (especially in areas receiving fog).

The CAM method has the distinct advantage of allowing plants to survive in harsh, droughty environments. Is there a trade-off? As with most things in life, the answer is yes. Have you ever wondered why tillandsias do not grow as large or fast as the typical garden plant?

One reason is that, all other things being equal, the CAM process is not as good at producing sugars over a given amount of time. Xeric tillandsias are less efficient at photosynthesis, so they grow more slowly.

Light and the Cultivation of Tillandsias

The question of how much light to give tillandsias is difficult because so many factors determine the answer. Nonetheless, there are some basic rules. First and foremost, unlike many houseplants, xeric tillandsias must receive strong, bright illumination. They are capable of withstanding long periods with poor or even no light, but eventually they must receive it in sufficient quantity. Most of them will benefit from light even brighter than what one would normally think. Remember, higher brightness levels mean higher temperatures, a corresponding increase in the rate of transpiration, and a concomitant increase in the metabolic rate. As a result, the plants will need increased access to water.

The position of the plant on the continuum between highly xeric and highly mesic indicates a great deal about its cultural preferences. Questions to ask yourself are, "Is the plant xeric—that is, gray, hard and/or stiff-leaved? Or mesic—that is green and soft-leaved? Is the plant small or large? Are the leaves thick or thin?"

Smaller, xeric species have less total leaf mass than larger plants. A smaller size means less total water volume. Intense sunlight will cause one of the minature species, *Tillandsia usneoides* Linnaeus, or "Spanish moss," to dehydrate more quickly than other, larger atmospherics, such as *T. xerographica*. With high levels of PAR, one should take care to watch the smaller plants for signs of dehydration; they should be watered accordingly.

If the leaves are thick, much of the light impinging on the plant does not reach the photosynthetic cells because it is absorbed by the water-storage tissues just below the epidermis. Most of the xeric species have thick, succulent leaves expressly for water storage. In combination with reflective trichomes and the less efficient CAM method of photosythesis, one can readily understand the need for high light exposure.

At the same time, however, each species has its own LIGHT SATURATION POINT. This is the level at which a plant cannot produce carbohydrate sugars at a greater rate, no matter how much more light it receives. In addition, plants receiving too much PAR can suffer from an effect called PHOTOBLEACHING. Photobleaching can damage and/or destroy chlorophyll pigment. Trichomes help to protect plants from this, but mesic species are more vulnerable because of sparse trichome distribution. When leaves bleach more yellow, it may be a sign that the plant is receiving too much light. If carried too far, the chlorophyll will be permanently damaged and/or the leaves may burn.

Likewise, there are limits as to how little light a plant can tolerate. Without enough light, the plant cannot produce the sugar necessary to maintain an adequate energy supply. As one would expect, some *Tillandsia* species have a remarkable capacity to adapt to lower light levels, especially those with thin leaves and/or few trichomes. The light intensity at which no net photosynthesis takes place is called the LIGHT COMPENSATION POINT (LCP). At this level of exposure, sufficient sugar is produced to just compensate for the quantity of carbohydrate respired over the same time period. A plant at this stage is "on hold," just balancing respiration. At light levels below the light compensation point, the plant is losing ground and eventually will exhaust all energy reserves. This period may be from a few weeks to a few months, depending on the species, the initial strength of the plant when put in the stressful situation, and how much the PAR level is below the LCP.

Despite the fact that each species is best adapted to a specific environment, most are flexible enough to survive in other situations that are quite different. This advantage allows growers to cultivate a large number of different species within a narrow range of cultural conditions. The question, "How much light should my plants get?," is easy to answer for the neophyte. Give them bright light outside and very bright light inside. If the plants are greener than average, provide a little less light. If they are stiffer and/or grayer, give them a little more light. Start from there and use your own observations to decide. If you want to grow the plants to their maximum potential, experimentation, common sense, and experience will point the way.

Fig. 6.22 Unidentified *Tillandsia* growing as an epiphyte on cactus in northern Peru.

Tillandsia Blooms and Vegetative Offsets

To quote Prof. Werner Rauh (1979), renowned bromeliad enthusiast, scientist, collector, and author from the University of Heidelberg in West Germany, "Every bromeliad grower, of course, likes to see his plants bloom. But in comparison to the generally plain leaves of orchids, there are among the bromeliads many species which even in their nonblooming state are the pride of any collection and evoke the delight of their owner."

Tillandsias certainly qualify on that score, but even though remarkable and intriguing, they are truly spectacular when in bloom. The brilliant colors range across the visible spectrum for variety. In addition, each inflorescence is a wonderful complement to the plant it graces. A quick look at the photographs in this book will attest to that. Their beauty points once again to the incomparable design of Nature. And yet, one should be aware that the beautiful blooms, as in the fascinating shapes of the plants themselves, have a definite purpose. Their size, shape, color, and blooming cycle all serve a very specific function—to attract a particular bird, insect, or other animal for pollination. Without the ability to attract animal pollinators, the plants would not exist.

This section will explore a most important aspect of the plants' life cycle—the blooming and reproductive season. Why do some tillandsias bloom earlier in the year than others? What triggers this process? Why do some species bloom for long periods of time, and others for only a few days? Why are the floral bracts and flower petals shaped and colored differently among different species—even within populations growing in the same tree? Why are some blooms fragrant and why do some plants become red when they are ready to bloom?

Living, primitive, nonflowering plants include, among others, algae, mosses, fungi, ferns, cycads (sago palm), and coniferous trees. Flowering plants, termed ANGIOSPERMS, made their appearance on earth 120-125 million years ago and quicky dominated the world's vegetation. Of the 300,000 or so green plant species on earth today, fully 250,000 produce flowers.

The Inflorescence

The *Tillandsia* inflorescence consists of a number of distinct parts. Many species bear the flower on a distinct SCAPE (stem that elevates or exserts the flower head out from the leaves). The usually abbreviated leaves positioned on the scape are often brightly colored, and are termed SCAPE BRACTS.

The spikes that hold the actual flowers are known as BLOOM SPIKES. Species with multiple or branched bloom spikes have a leaf at the base of each spike called a PRIMARY BRACT, and these bracts are often brightly colored. FLORAL BRACTS, are the generally colorful, much reduced and modified leaves on the bloom spikes from within which the flower emerges. In *Tillandsia*, floral bracts initially cover the entire developing flower. They serve to protect the floral apparatus as it grows and later, if colorful, help to attract animal pollinators, especially hummingbirds. Interestingly, floral bracts are sometimes a different color than the scape and primary bracts.

The enlarged stem base upon which floral appendages are attached is called the RECEPTACLE, and it bears four types of appendages. Attached lowermost on the receptacle is the CALYX (KAY-lix), which is made up of three SEPALS (SEE-pulls) that serve to protect the more apically

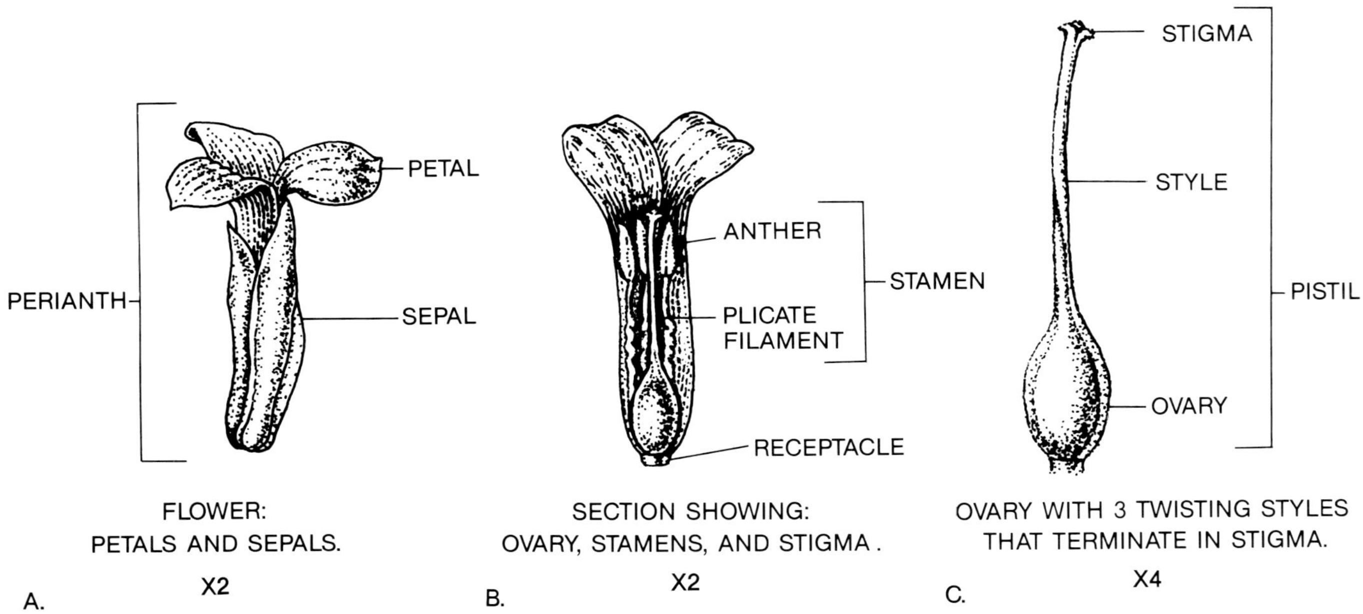

Fig. 6.23 *Tillandsia aeranthos* (X2).

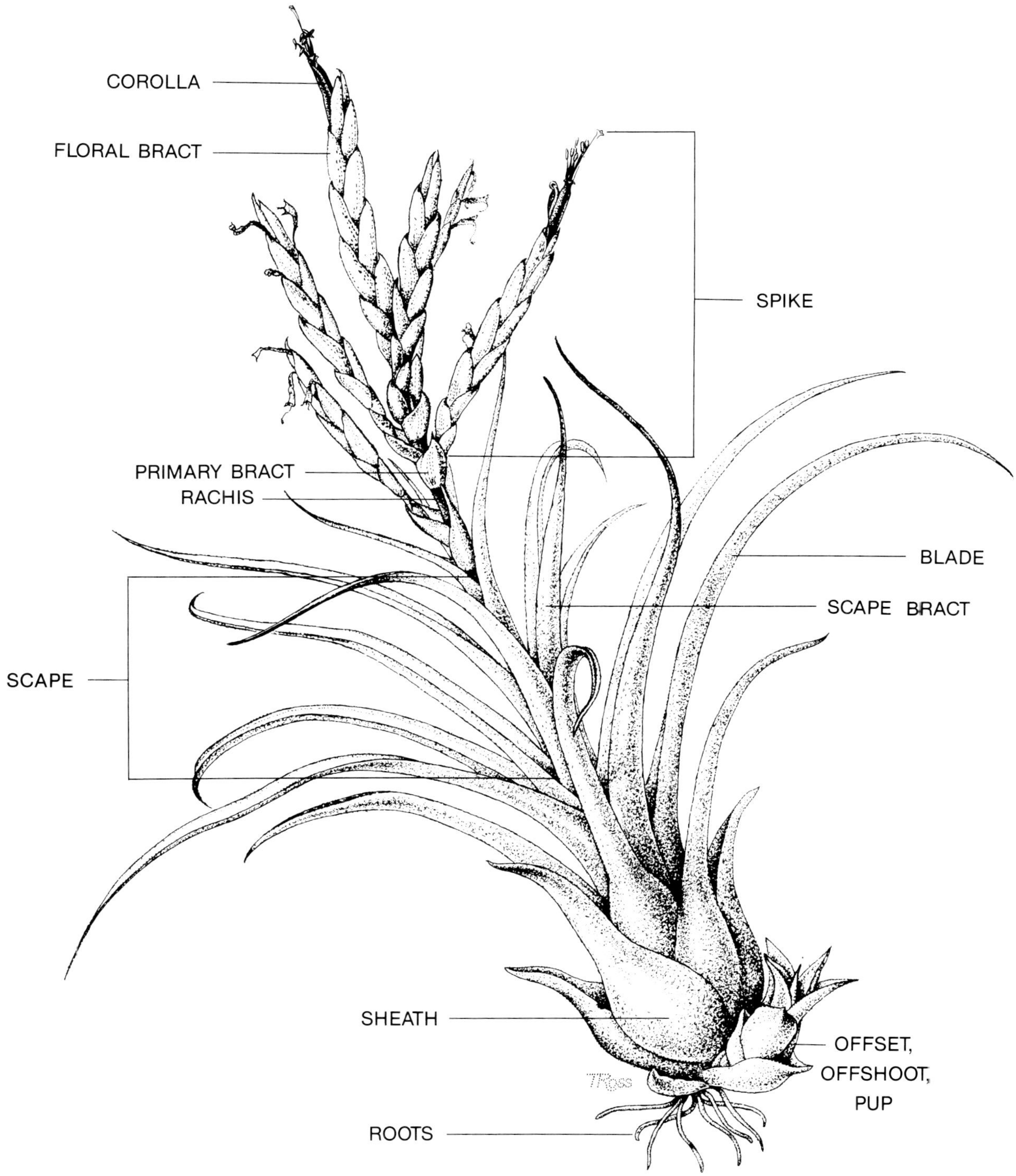

Tillandsia X wisdomiana

Fig. 6.24

inserted parts of the flower, the PETALS, STAMENS, and PISTIL. The calyx members are green and rather tough, and often hidden by the floral bracts. The COROLLA is inserted above and inside the caylx and consists of three separate petals that are often brightly colored.

A number of flowering plants, such as oaks and grasses, are wind pollinated, but most angiosperms, including tillandsias, attract animals for the task. Hummingbirds, moths, butterflies, bees, moths, and even bats pollinate tillandsias.

Scientists have deduced trends in floral traits using information gathered from studies of many angiosperm families. Certain floral shapes or characteristics have traditionally been associated with specific pollinators. For instance, flowers pollinated by birds are generally DIURNAL, which means that they have a daytime ANTHESIS (flowers fertile or receptive to pollination), and display vivid parrotlike colors. According to Gardner (1982), the flowers of ORNITHOPHILOUS (ore-ni-THAW-phil-us; bird-pollinated) species are often tubular, with little or no lip to the petals. If a lip is present, it curves back. The often violet flowers are mechancially strong, having a stout corolla tube and stiff filaments with fixed anthers.

Faegri and Van Der Pijl (1979) stated that PHALAENOPHILOUS (fah-lye-NAW-phil-us; moth-pollinated) flowers are similar to ornithophilous flowers, but that anthesis is generally NOCTURNAL and the flowers are often sweetly fragrant. Corolla colors are usually white, green, or sometimes pale red. Corolla tubes are typically more narrow and not as mechanically strong as those of ornithophilous flowers. Flowers pollinated by butterflies are vividly colored and similar to the above and tend to be diurnal and weakly scented.

MELITTOPHILOUS (me-lit-TAW-phil-us; bee-pollinated) flowers are diurnal and must be mechanically strong and have

Fig. 6.25 Bright red dots of *Tillandsia bulbosa* pepper Rio Dulce river bank in Guatemala, making an attractive target for foraging hummingbirds.

Fig. 6.26 Closeup of an ageotropic *Tillandsia bulbosa* in bloom along Rio Dulce river bank. Note myriad of bromeliad seedlings.

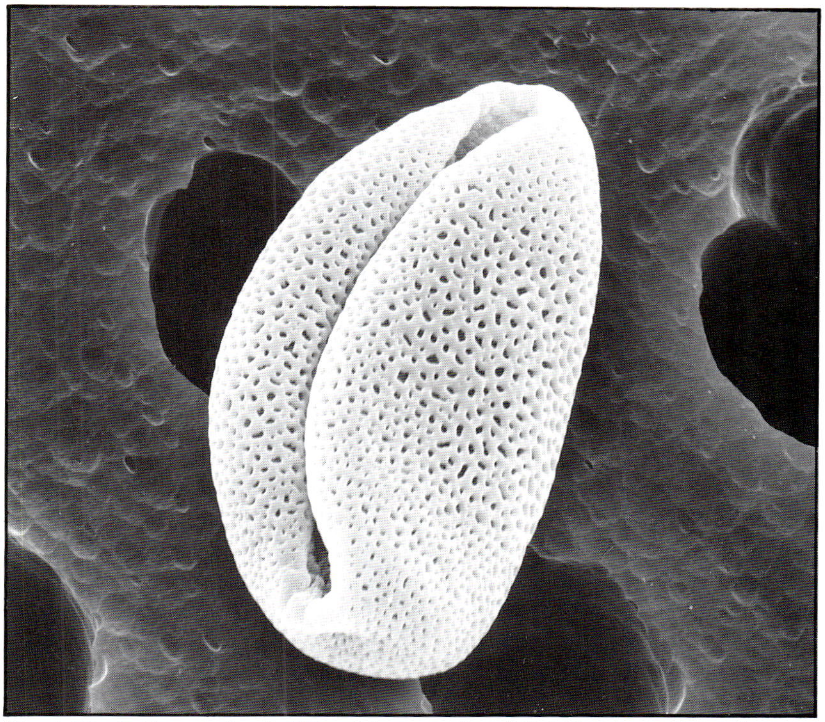

Fig. 6.27 Pollen grain of *Tillandsia aeranthos* (X1,300; photo Brown).

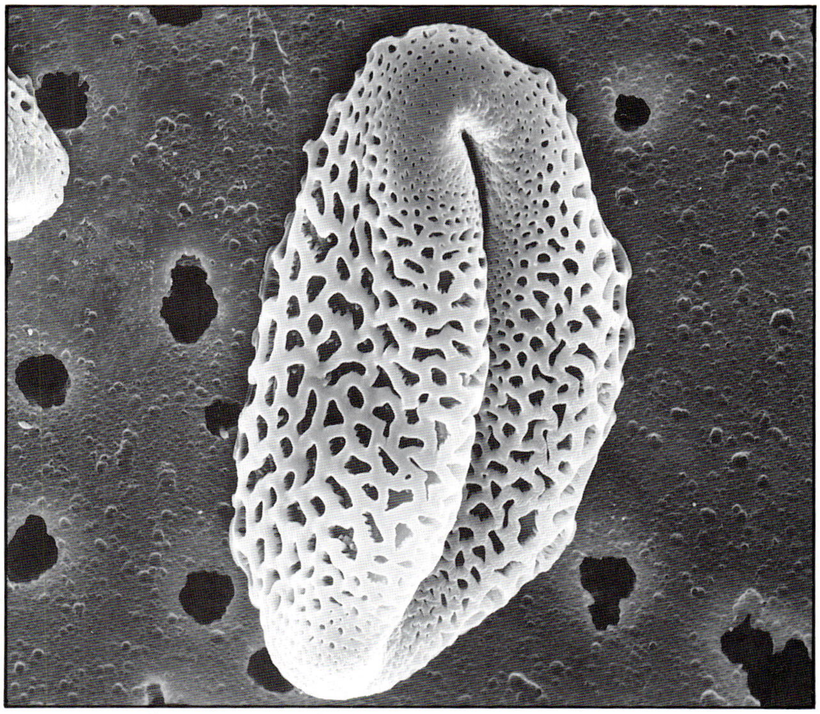

Fig. 6.28 Pollen grain of *Tillandsia jucunda* (X2,100; photo Brown).

petal lobes large enough for the bee to land. Colors may be bright, with yellow and blue being the most frequent. Odor is not usually strong.

Some of the bird-pollinated species, such as *Tillandsia ionantha, T. brachycaulos,* and *T. bulbosa* actually become red when in bloom, and hummingbirds are drawn to that color. Fragrant blooming species such as *T. straminea, T. duratii,* and *T. xiphioides* are moth-pollinated. Fragrant species, like *T. dodsonii* L.B. Smith and *T. narthecoides* Presl have large white flowers, bloom at night, and are probably pollinated by nocturnal moths. Our knowledge of specific *Tillandsia* pollination biology or breeding systems is sketchy, and much research remains to be done in this field.

Sepals and petals occur in WHORLS—that is, they are arranged in circular fashion on the receptacle. The calyx and the corolla together are called the PERIANTH (PEAR-ee-anth).

Tillandsias grown from seed are the product of a sexual union between a SPERM and EGG. These sex cells are called GAMETES (GA-metes). The seedling grown to adult will exhibit a new, unique combination of the genes. This is important in an evolutionary sense because with a more dynamic gene pool, adaptation to new growing

Fig. 6.29 Bloom of *Tillandsia ionantha v. scaposa* (X2).

conditions can be faster and more varied.

STAMENS, the male appendages of the flower, are attached to the receptacle just above the perianth. They are grouped in two whorls of three. Each of the six stamens consists of a long slender FILAMENT at the end of which is the POLLEN bearing portion called the ANTHER. The anthers and pollen grains of species in subgenus *Tillandsia* are easily seen because they are visibly exserted above the petals. The anthers and pollen grains are usually bright yellow. An anther contains four pollen sacks that are filled with the grains.

Pollen grains, of which there are two basic types in *Tillandsia*, are shaped like partially inflated footballs. The honeycomblike covering on the grain is called the EXINE. In one *Tillandsia* group, the pores of the exine are approximately the same size. In the other group, the pores are larger in the center and get smaller at the ends of the grain. The exine serves the structural function of protecting the two cells inside the grain.

The pollen grains (Figs. 6.27 & 6.28) have been washed in a solution that removes the glutinous carbohydrate, sugar, and protein substance from the grain exterior. The pores of the exine hold this substance to the grain. The characteristic makeup of the substance is crucial to the pollination process. Receptors in the stigma (Figs. 6.30 & 6.31) accept or reject the pollen based on the compatibility of the ingredients in this coating. If the stigma is compatible with the coating on the grain, the pollination process begins.

Each pollen grain produces two haploid nuclei. One of these two cells becomes the POLLEN TUBE. The slit, or GERM PORE, in the pollen grain allows the pollen tube to emerge. The pollen tube tunnels its way down the style of the stigma. The other cell inside the grain, the GENERATIVE NUCLEUS, splits and becomes two sperm cells. These cells make their way down the pollen tube where they

Fig. 6.30 Stigma of *Tillandsia fasciculata* (X90; photo Brown).

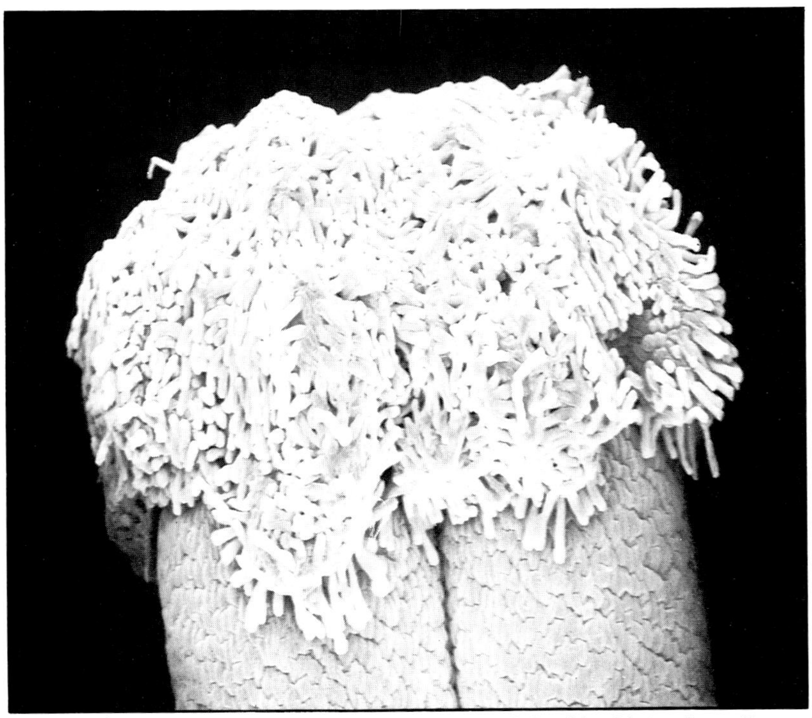

Fig. 6.31 Stigma of *Tillandsia venusta* Mez & Werckle (X90; photo Brown).

may eventually make contact with an ovary. The fertilized egg, or zygote, then develops into a seed.

In the center of the flower is the female sexual apparatus, the pistil. This is the most complex of the flower's appendages. It includes the swollen basal part called the OVARY. Inside the ovary are three chambers, each of which contains one or more OVULES. Ovules are structures in which the female haploid gametes develop. Should the female gamete, or egg, become fertilized, the ovule develops into seed. The long, slender STYLE extends from the top of the ovary and terminates in the STIGMA. The stigma shape for each species is different, and Brown and Gilmartin (1984) have found that stigmas for all bromeliads are of three basic shapes.

Vegetative Offsets

Most *Tillandsia* shoots are typically DETERMINATE, that is, they grow for a predetermined length of time, then bloom just once. However, most species produce numerous offsets. A few species such as *T. multicaulis* Steudel and *T. complanata* Bentham are INDETERMINATE, i.e. the same shoot continues to grow and flower season after season. Also, these two species are unusual in that they produce multiple blooms. Occasionally, these plants will also produce offshoots.

Some species, however, are MONOCARPIC. *Tillandsia guatemalensis* L.B. Smith, *T. prodigiosa* Baker, and *T. eizii* L.B. Smith bloom once, develop and scatter seed, then die. They produce no offsets.

In stark contrast to the manner in which many *Tillandsia* species produce offsets that mature in a year or two, a sad fact for the collector is that seed-grown plants often take five to eight years to produce a flowering specimen—a long time to maintain one's patience and interest. For those who do, there is great inherent satisfaction.

Contrary to popular myth, the parent plant of many determinate shoot species will exhibit vigor for two or even three generations before disappearing completely. *Tillandsia caput-medusae, T. seleriana, T. ionantha,* and *T. bulbosa* are examples. One may easily determine the oldest plants in a clump by the degree to which the past bloom individuals have died back or have been overgrown by younger plants.

Since numerous offsets are normally produced in most species, just a few generations are needed to establish a thriving clump of plants that bloom with vivid color. This is the goal of many collectors. Vegetative offsets rarely differ genetically since asexual reproduction is involved. They can and often do change in physical morphology from their parent because environmental conditions play such a large role in the plants' appearance.

Many of the atmospheric *Tillandsia* species mature and bloom on a yearly or biannual basis; the blooming cycle itself is from just a few weeks to several months. *Tillandsia* shoots that take more than one season to reach maturity bloom for longer periods of time. *Tillandsia rothii,* which matures in two to four years, may have a blooming cycle of a year or more.*

In nature, the largest proportion of *Tillandsia* species bloom in winter. The reason for winter blooming is that tillandsias grow from pup to maturity when the days are long and warm. In the fall, with shorter and cooler days, leaf growth halts and the apical meristem changes from producing primordial leaves to the irreversible sexual process of developing a fertile shoot—the inflorescence. The ensuing months are devoted to this, after which the effort in most tillandsias goes into the production of offshoots and/or seeds.

Flowering is probably faithful to timing in natural habitats, and wet or dry weather is most likely the selective force that determines that timing. Day length is used to time flowering to the correct season, except for equatorial regions where day length does not vary. Seeds are usually released just before the onset of wet weather. Flowering is presumably synchronized to coincide with the availability of appropriate pollinators for the particular species.

In cultivation, a plant must be strong to produce a large, healthy inflorescence, because great energy is used in its production. This is a major reason tillandsias usually will not bloom when placed indoors. Most indoor situations just do not have enough light to allow the plant to build up sufficient nutrient and energy reserves for sexual reproduction. In addition, tillandsias grown indoors may not receive the correct stimulus, such as day length, cold spell, and so forth. A desiccated *Tillandsia* will have a difficult time producing a fertile inflorescence because water, like light, is an essential source of life. Most collectors cultivate their plants outdoors when possible, although many bring blooming specimens indoors to enjoy them more fully.

**See* Comparison Chart.

Literature Cited:

Brown, G. & Gilmartin, A.J. 1984. Stigma Structure and Variation in Bromeliaceae—Neglected Taxonomic Characters. Brittonia, 36(4), pgs. 364-374.

Faegri, K. & L. Van Der Pijl. 1979. *The principles of Pollination Ecology.* 3rd revised edition. N.Y. Pergamon Press.

Gardner, C.S. 1982. *A Systematic Study of Tillandsia Subgenus Tillandsia.* Ph.D. dissertation. College Station. Texas A & M University, pg. 79.

Rauh, W. 1979. *Bromeliads.* Dorset, England. Blanford Press, Limited, pg. 37.

Tillandsia Seeds

Each species of *Tillandsia* is capable of producing from a few dozen to thousands of seeds in a reproductive season. Every pollinated flower produces one three-chambered seed pod. The number of seed capsules produced by a plant varies greatly in the genus *Tillandsia*. *Tillandsia albertiana* produces only one seed pod, while monocarpic species such as *T. guatemalensis* are capable of producing hundreds.

All *Tillandsia* seeds bear a tuft of COMA HAIRS that represent an outgrowth of the seed coat. These coma hairs, although greatly compressed within the fruit, or capsule, still occupy most of the inner cavity. The seeds are grouped at the apical end of the pod. The time it takes for the seeds to ripen varies from three months to two years, depending on the species.

Gradually, the pod desiccates and the walls of the three chambers that make up the capsule slowly shrink, eventually reaching a point where they split apart. Because of the mechanics involved in the shrinking process, when DEHISCENCE occurs the outer wall of each chamber curls back, providing a large opening from which the seed can escape (Figs. 6.33 & 6.34). Seeds are subsequently scattered, usually by wind. The coma hair structure, or "parachute" as it has come to be known, resembles a fully opened parachute.

The structure and coarseness of the coma hairs serve a twofold purpose. The shape of the parachute is highly efficient for airborne movement. It is light, well designed, and balanced to carry the light seed long distances. Also, the coarse, barbed hairs and their radial spread at the top are ideal to catch and hold on to tree bark, rock, or other support.

Once contact is made, other hairs quickly become entangled, providing attachment. Here, with the ideal conditions of warmth and brief thundershowers interspersed with periods of sunlight, the seeds will germinate and grow,

Fig. 6.32 Unripe seed pods of *Tillandsia stricta*.

Fig. 6.33 Open *Tillandsia* pod showing 3 carpel shells peeled back.

Fig. 6.34 Dehisced seeds of *Tillandsia tricolor*. Note dark color of interior skin of carpel.

Fig. 6.35 Seeds of *Tillandsia ionantha* ready for airborne flight.

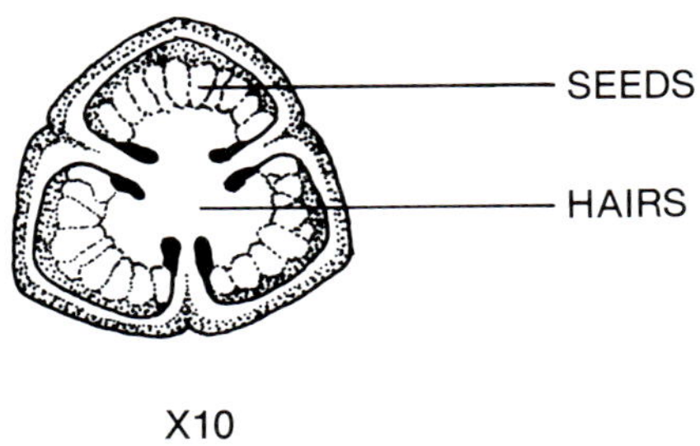

Fig. 6.36 Cross section of *Tillandsia aeranthos* seed capsule. Note placement of seed.

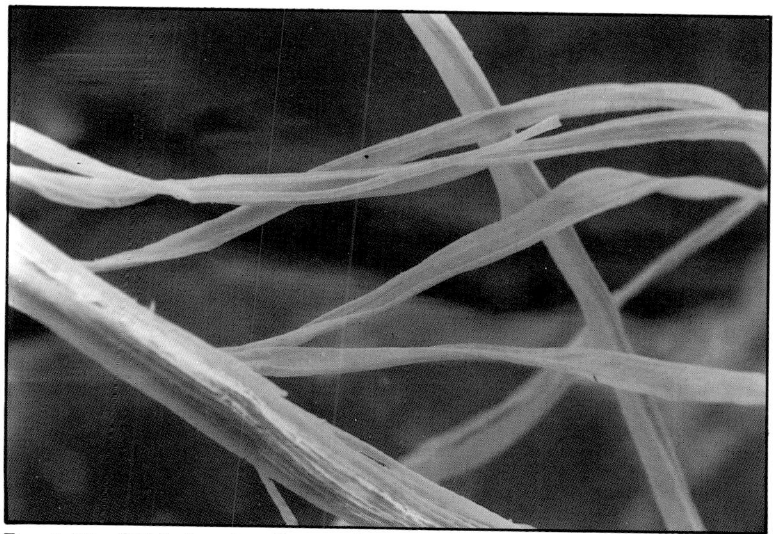

Fig. 6.37 SEM showing flattened coma of *Tillandsia lepidosepala* L.B. Smith (X240; photo Brown).

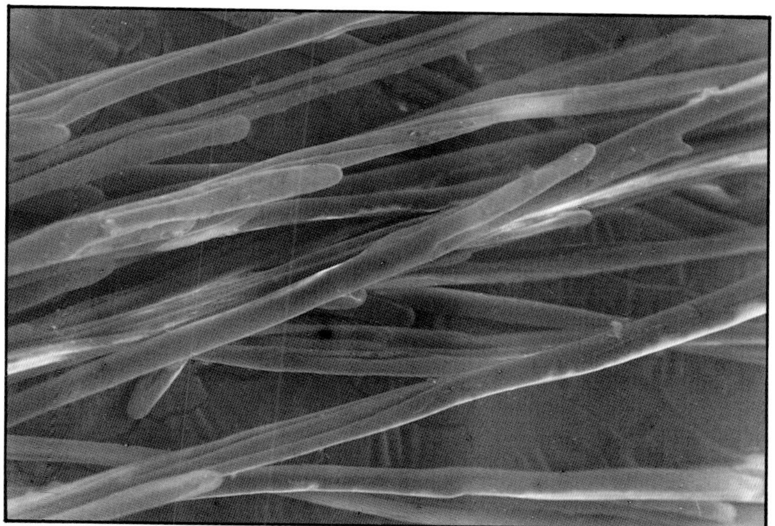

Fig. 6.38 SEM showing barbs on individual coma of *Tillandsia balbisiana* (photo Brown).

albeit very slowly for the first two years or so.

In their natural habitat, tillandsias often time the release of their seeds to just before the onset of the rainy season. Some dry weather is required to ensure sufficient time for dispersal to new locations via air currents. If wetted too soon, many seeds would be immobilized on the inflorescence. This would allow for the survival and maturation of just a few plants and would not be an effective means for increasing the numbers and range of a species.

At the same time, however, when seeds do become wet and the germination process has begun, the new seedlings are sensitive to drought relative to adult plants. They are subject to rapid dehydration and can die quite easily, although many *Tillandsia* seedlings are remarkably drought tolerant compared to seedlings of most other plants. Seedlings need constant humidity until the first leaves are produced and the plant is able to survive dry spells.

Although species sharing the same habitat may disperse seeds at about the same time, flowering need not, and usually is not, synchronized. This is because flowers and fruits of different species mature at varying rates.

The manner in which *Tillandsia* flowers are pollinated is rather primitive compared to a few other plant families, such as orchids, but seed dispersal is very specialized in *Tillandsia*. Many species have radiated hundreds or even thousands of miles, and although much of this dispersal has probably been accomplished by birds, much also has been the result of the airborne flight of naked seeds.

Tillandsia Roots

Tillandsias develop an extensive holdfast root system from the early seedling through juvenile stages. Their function is to take over plant attachment from the coma hairs. In nature, xeric species produce fewer roots as they mature. Some species, such as *Tillandsia duratii*, produce virtually no roots at all once the early growth stage has been passed. In contrast, mesic, terrestrial species produce more roots and use them for moisture and nutrient uptake.

In cultivation, the amount and vigor of root development depends a great deal on the conditions under which the plants are grown. Many xeric species, such as *Tillandsia juncea*, will grow extensive root systems if potted in a loose, fast-draining soil mix. These roots will be delicate and have many branching hairs that will absorb moisture and nutrients. At the same time, mesic species such as *T. lucida* Baker develop virtually no roots when grown epiphytically under arid conditions. The growing environment plays an important role as to whether or not a plant will develop roots. Moisture and constant high humidity are the keys to vigorous root growth.

The live growing tips of *Tillandsia* roots are yellow (Fig. 6.40). When grown with the plant base near a mounting medium, the roots gradually will make contact with it, crawling across the surface

Fig. 6.39 Tremendous root development by *Tillandsia latifolia* 'Enano.'

Fig. 6.40 Developing root tip of *Tillandsia straminea*.

Fig. 6.41 Striking windmill pattern created by extensive roots and leaves of *Tillandsia lorentziana*.

Fig. 6.42 Freaky, abstract, yet proportional root development of *Tillandsia edithae*.

Fig. 6.43 Blooming *Tillandsia tectorum* on exotic cholla (cactus). Roots have formed an interesting addition by taking advantage of the natural openings in skeleton. Note yellow, developing root tips.

Fig. 6.44 Clump of *Tillandsia aeranthos.*

Fig. 6.45 Detail of stem and roots after part of the clump has been gently but firmly pulled apart.

and attaching the plant in a manner similar to ivy. As the roots develop, all but the yellow tip of the epidermal surface soon dies.

Roots sometimes grow in an upward direction, growing among and attaching themselves to the leaves. Roots of epiphytic and saxicolous species are THIGMOTROPIC, which means they grow against adjacent bark or rock regardless of their orientation relative to gravity. When grown without a substrate in close proximity, such as when hung from a piece of wire, the roots will often grow geotropically, that is toward the source of gravity.

Tillandsia roots are adventitious; they develop from the plant base or along the stem rather than from a tap root. As the stems of caulescent species grow, new roots are produced further up the stem. The older part of the stem dies, along with the roots.

One can easliy remove roots considered unsightly by trimming them with scissors. Roots that have attached themselves to the leaves can be peeled and removed if desired. Roots do not have the strength to penetrate wood or any other hard substance. They do penetrate crevices and openings (Fig. 6.43).

In the garden or greenhouse, most xeric species will produce root systems during the growing season if cultivated under humid conditions. In a prolonged dry spell, the root tips are the first part of the plant to die. Roots grow best with high humidity. They are not necessary for good shoot growth, but when present, are an indicator of favorable growing conditions.

Foliar Trichomes

The white, frosty coating on xeric tillandsias is actually a continuous layer of thousands of tiny light-reflective, absorbant HAIRS or SCALES. They are the FOLIAR TRICHOMES. These epidermal appendages, which are morphologically unique for most species, are among the most interesting and important mechanisms by which atmospherics cope with harsh environments. An understanding of the basic trichome structure will allow the reader to appreciate how they act as a front line of defense in protecting the plant from the sun's excess radiant energy (PAR), while at the same time performing as a front line of offense in the battle to secure the plant's share of precious water during droughty periods.

Tillandsia trichomes are made up of groupings of dead cells. At the center of each trichome is a flat plate of cells termed the SHIELD, which consists of a grouping of four concentric CENTRAL DISC CELLS. Depending on the species, further concentric series of cells surround the central disc in numerical groupings of 8, 16, 32, or 64. These are called RING CELLS. The outer circle of ring cells is attached to the most visible part of the trichome apparatus, the WING.

Trichomes may densely cover the epidermis or be widely scattered. The wing of an individual trichome may be almost symmetrical, as in *Tillandsia xerographica*, or significantly larger on one side, as in *T. tectorum*. When dry, the wings are separate from and above the surface of the epidermis. The wings are appressed-spreading in some species such as *T. utriculata*. When dense and overlapping, appressed scales bestow a velvety look and feel to the leaves. In contrast, the wings of species such as *T. ionantha* rise up at an almost perpendicular angle from the surface, giving the leaves a coarse appearance. The wings of most species are flexible, but those of *T. bulbosa* and *T. butzii* are inflexible.

The central disc cells have upper and lower surfaces or walls. The upper walls are thick. When completely dry, they sink into the appropriately sculptured depressions formed by the lower walls. This snug fit reduces the volume of the LUMINA (cavity within the cells) to a small fraction of their maximally distended state.

In the collapsed configuration, the cell walls of the four central disc cells form a plug preventing the escape of moisture from the leaf interior via the STALK CELLS. Without the trichome shield, the stalk cells would act as a wick, pulling moisture from the leaf and allowing it to evaporate into the atmosphere. As an analogy, imagine a water-filled bottle with a piece of absorbent cotton placed in the neck.

Stalk cells are beneath and perpendicular to the shield. The uppermost stalk cell is termed the DOME CELL. Because the shield and stalk cells together resemble an umbrella, the trichome apparatus is sometimes referred to as a PELTATE SCALE.

When the cells of the leaf are the least bit dehydrated, they will absorb water if it is made available. Water reaching the dome cell flows through the stalk cells into parenchyma cells of the hypodermis, where it is stored for future use. Therefore, shield cells create a one way valve effect, allowing the plant to absorb moisture quickly on one hand, and yet serving to retard water loss during dry weather. Fully turgid cells will not absorb more water.

Walls of the trichome cells are composed of organic (carbon-based) molecules consisting primarily of CELLULOSE and some PECTIN. Cellulose is a long, flexible, unbranched molecule consisting of repeated units of smaller GLUCOSE molecules. Pectin is a branched POLYMER (a long chain of molecules, composed by the repetition of relatively simple molecules).

As a trichome cell develops, cellulose is manufactured in the cell protoplast, but the protoplast dies soon after it matures, leaving walls composed of the cellulose molecules. Together, as a group, the molecules are fibrous in texture. They attract water by capillary movement (like a paper towel) and possess the ability to hold relatively large quantities of either air or water. When water comes into contact with a trichome, it is quickly absorbed. The thick upper walls of the central disc and ring cells rise, while the flexible side walls straighten.

In contrast, the upper walls of the outer ring cells are thin and flexible while those of the lower walls are thick and rigid. The somewhat convoluted shape of the outermost ring cells is important because of their relationship with the wing cells.

When trichomes become saturated with moisture, there is a torsion due to the interaction of cellulose fibers and water. The lower cell walls of the outer ring cells straighten as they take up water. The thin upper wall stretches during this process, and the wing is drawn down against the epidermis. All of these movements occur very rapidly.

Once a moisture source has been removed, the leaf surface and trichomes begin to dry. Trichomes dry rapidly because they lack protoplasts and waxy cell walls common to epidermal cells. The plant quickly regains its white fuzzy appearance, the central disk and ring cells collapse, and the wings flex away from the epidermis.

Trichomes are white because, when dry, they are filled with air that scatters sunlight. When wetted, water forces air out of the trichomes, rendering them transparent, so that the green chlorophyll within the mesophyllic tissue becomes visible.

The trichomes of jungle or cloud forest, tank-type, mesic species, such as *Tillandsia punctulata*, are very small and widely dispersed. They do not play a significant part in moisture absorption or nutrient uptake because sufficient water and nutrient, rich detritus collect in the leaf axils (bases). In addition, trichomes are not needed to reflect excess PAR, because light levels are generally low.

Water movement through *Tillandsia* trichome.

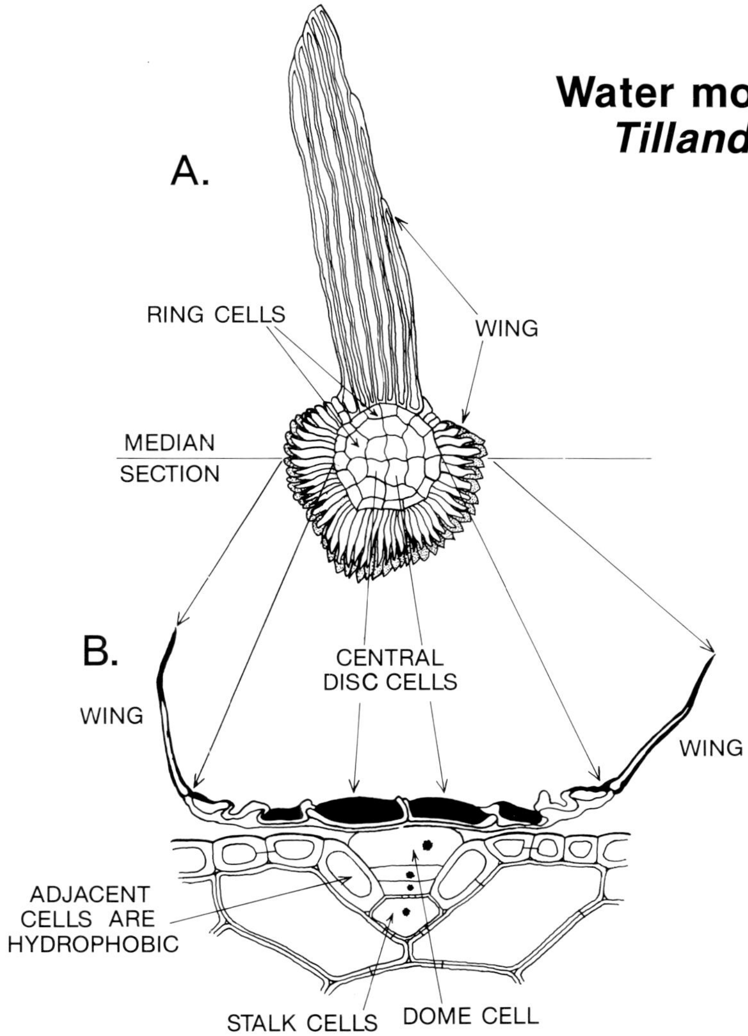

A. Trichome of *Tillandsia crocata.* Wing cells of xeric tillandsias, such as *Tillandsia crocata, T. tectorum,* and *T. plumosa,* stand up perpendicular to the epidermis in the dry state.

B. Median section through a dry *Tillandsia* trichome. Note upward flexion of wings.

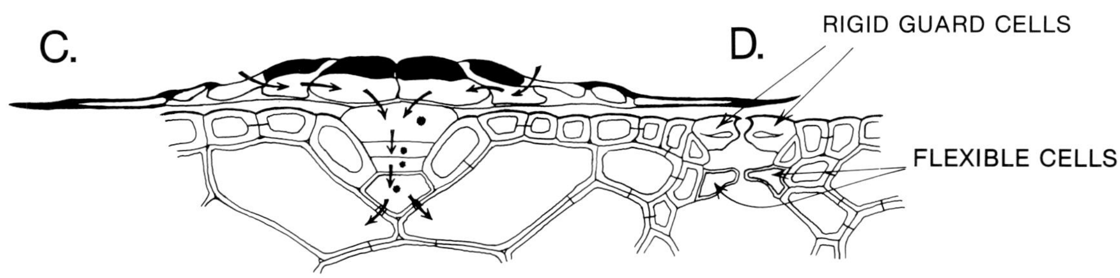

C. Same trichome when the surface is moist and water is moving through the hair into the leaf interior. Arrows depict the path of uptake.

D. A stoma in cross section.

Fig. 6.46 (redrawn from Benzing)

With the atmospherics, the trichome has evolved to its most sophisticated state in the Plant Kingdom. Xeric tillandsias would not be able to endure the harsh climate of the natural habitat without a confluent covering of trichomes.

Trichome Variations

In an overall sense, trichome variation among species is primarily an accommodation to the demands imposed by the environmental conditions that prevail in the plants' habitats and are not related to floral anatomy, which is the basis for classification. As interesting as this field of study would be, trichomes are not used to establish evolutionary (taxonomic) pathways.

Also, according to Tomlinson (1969), the trichomes on a single leaf may vary. Scales at the base of a leaf may differ from those that are more distal in features such as the number of stalk cells, the size of the scales, and the degree of overlapping of the trichome wings, as well as in the shape and degree of wall thickening of shield cells. Similar differences can occur among scales at the same level of a leaf but on opposite sides. This makes it difficult to synthesize an overall trichome design for an individual species.

However, certain patterns of trichome development and their purposes have been established. This section will discuss a few of these strategies and some of the theories concerning their functional significance.

The variety of *Tillandsia tectorum* that grows at high altitudes in the Andes of Ecuador is exposed to intense levels of solar radiation including dangerous ultraviolet wavelengths. Although actual rainfall is rare, during certain times of the year, *T. tectorum* is also subject to frequent soakings because of passing clouds. These are interspersed with periods of intense sunlight that easily could damage sensitive leaf tissue were it to be fully exposed.

As an adaptation, the plant has evolved a means to enhance evaporation between soakings: in order to 1) protect the leaf from excess PAR and 2) avoid suffocation. To enhance evaporation of water, trichome wings rise up quickly from the leaf as the drying process commences. This exposes the epidermis to maximum air movement. At the same time, attenuated trichomes scatter a large percentage of incident radiation. Many species possess this characteristic to a greater or lesser degree.

The coarse surface texture created by the trichome wings of *Tillandsia tectorum* may allow the plant to intercept airborne particulates more effectively. This matter could release mineral nutrients for absorption while lodged among the scales on wet leaf surfaces. This feature may be of importance in an oligotropic habitat.

In contrast, other species must develop ways to protect the plant from dry atmospheres, infrequent rainfall, and/or often intense desert heat. This is in addition to high levels of PAR.

In three locations in Guatemala, *Tillandsia xerographica* thrives in dry, droughty conditions. The summers are hot and the short rainy season occurs only in July and August, at which time it may rain lightly three or four times a week. Except for a few sporadic showers, the rest of the year is usually rainless, although morning dew from the frequent cool nights provides some moisture. Protection from water loss and heat buildup, as well as excess PAR, are primary concerns of this large-sized species.

Tillandsia xerographica has developed trichome wings that parallel the epidermis. They overlap and densely cover the leaves and stomata. The flattened wings reflect sunlight, but in addition they act as a barrier between the leaf surface and surrounding atmosphere. In this capacity they create a boundary layer of insulating air that fulfills the two purposes of keeping the leaf tissue cooler and slowing down the rate of transpiration.

The trichome is an amazingly effective apparatus, but it also has drawbacks. The rate of water loss is low because the trichomes block the free exchange of water in the leaf and the surrounding air into which it has a high potential to evaporate. But at the same time, the trichomes act as barriers to oxygen and carbon dioxide gas exchange.* This contributes to the relatively slow growth rate for this and similar species.

Another trichome arrangement is found in species such as *Tillandsia bulbosa* and *T. butzii*. These species have developed different trichomes on the adaxial (upper or inner) and abaxial (lower or outer) leaf surfaces. At first, *T. bulbosa* and *T. butzii* seem to be contradictions in terms of plant habitat preference and the type and density of their trichome distribution. Here are two species with dense layers of trichomes, yet they not only flourish in shady environments, they also are subject to substantial rainfall! Current theory by Benzing (1978) holds that these two species may have evolved from more xeric ancestors to live in the generally shaded and moist situation.

Interestingly, when the plants are dry, their narrow adaxial, or ventral, leaf surfaces reflect light just as a typical xeric species; but the more exposed abaxial, or dorsal, surfaces do not, particularly those of *Tillandsia bulbosa*. In addition, water droplets coming into contact with the greatly involuted dorsal surfaces do not spread out as they would on the upper leaf surface, but simply bead on the waxy leaf surface.

The abaxial leaf surfaces of these species are not as reflective as the adaxial surfaces for two reasons: the abaxial surface in the case of *Tillandsia bulbosa* does not have a dense trichome covering, and in both species, the trichomes are quite different from other xeric species. The wings of the scales are small and fixed in place, parallel to the leaf surface instead of having the "hinged-flexible joint" like those of other xeric species. Both the upper and lower

*See subsection on CAM Photosynthesis, pg. 192.

walls of the shield cells are more or less uniformly thickened and rigid. Additionally, the abaxial trichomes are transparent to sunlight. All of this promotes the penetration rather than the reflection of light.

According to Benzing, the likelihood that these species have evolved from more xeric ancestors is evidenced by the trichome structure on the adaxial surface of the leaves. Here they are typical for a xeric species—that is, packed densely, hydrophylic, and flexible. Because the adaxial surfaces are on the inside of the involuted leaf surface, they are not exposed very much and their significance is relegated chiefly to water intake.

At least one species, *Tillandsia tetrantha* v. *aurantiaca* (Grisebach) L.B. Smith, has large trichomes, even though it lives in a low-light, moist habitat. Gilmartin (1972) learned that *T. tetrantha* v. *aurantiaca* was found "...in a very wet cloud forest region at about 2,500 meters. The region where this variety was so common, however, was remarkable for the amount of wind.... The winds certainly appear to bring about somewhat xeric conditions in what might otherwise appear to be nearly hydric conditions by causing high rates of evapotranspiration."

The near constant high wind levels create a significant drying effect that has been overcome by the trichomes. Tillandsias with dense layers of trichomes usually prefer high levels of light exposure because the trichomes scatter a large percentage of the light radiation that strikes them. *Tillandsia tetrantha* v. *aurantiaca* would be an exception.

Virtually all plants have some type of trichome, but it can safely be said that no other plants on earth have developed the use of the trichome to the extent of the xeric tillandsias. To quote Benzing (1978), "...to simply state that the foliar trichome found in Tillandsioideae functions in moisture and salt absorption and light reflectance is an oversimplification, and obscures the reality that these organs produce a multitude of additional effects which significantly influence the physiology of these plants and thereby their habitat preference."

Although Benzing and others have made remarkable strides in our understanding of *Tillandsia* biology in recent years, much of what is written about trichome function remains conjectural.

Literature Cited:

Benzing, Seeman, & Renfrow. 1978 The foliar epidermis in Tillandsioideae (Bromeliaceae) and its role in habitat selection: Amer. J. Bot. 65(3): 359-365.

Gilmartin. A. J. 1972. *The Bromeliaceae of Ecuador.* Leutershausen, West Germany. J. Cramer, pg. 6.

Tomlinson, P.B. 1969. *Anatomy of the Monocotyledons.* London, England. Oxford University Press, Ely House, pg. 218.

SCANNING ELECTRON MICROSCOPE (SEM) PHOTOGRAPHS OF *TILLANDSIA* TRICHOMES

In order to show some of the fascinating differences among the scales of various species, two scanning electron microscope (SEM's) photographs per *Tillandsia* are featured on the following pages. One is from the abaxial leaf surface, the other from the adaxial. The trichomes sometimes vary greatly from one side of the leaf to the other. Some photos show the detail in an individual scale while others are less magnified and present an overall view of the epidermis.

A few of the many questions you may want to consider are: how many series of ring cells are there? How many wing cells surround the ring cells? And are the wings coarse or appressed and to what degree? Trichome cell boundaries are sometimes difficult to distinguish, which makes definite answers difficult to ascertain.

As you view the the SEM's, realize that every facet of the trichome has its purpose. Botanists and others are just scratching the surface of inquiry at this time.

Tillandsia aeranthos
adaxial X450
abaxial X180

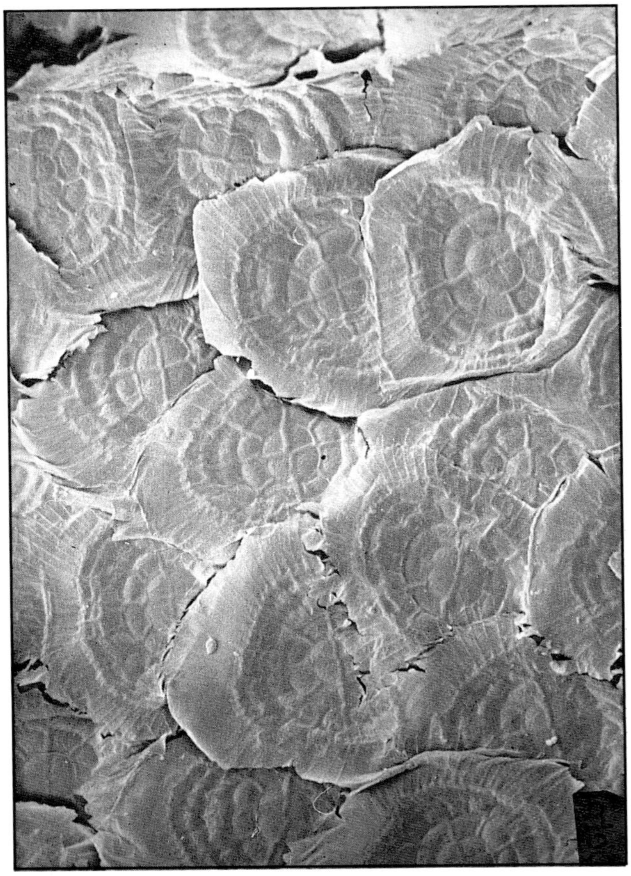

Tillandsia albertiana
adaxial X190
abaxial X380

Tillandsia argentea
adaxial X180
abaxial X390

Tillandsia argentina
adaxial X350
abaxial X180

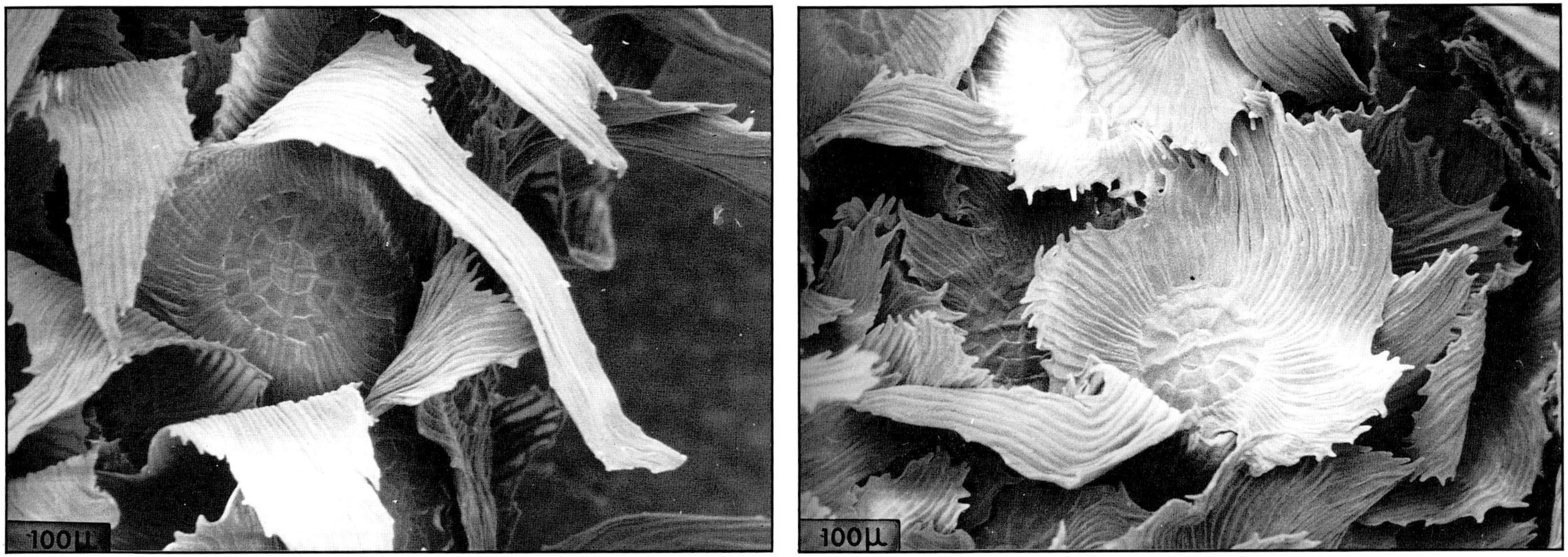

Tillandsia atroviridipetala

abaxial X 180 adaxial X180

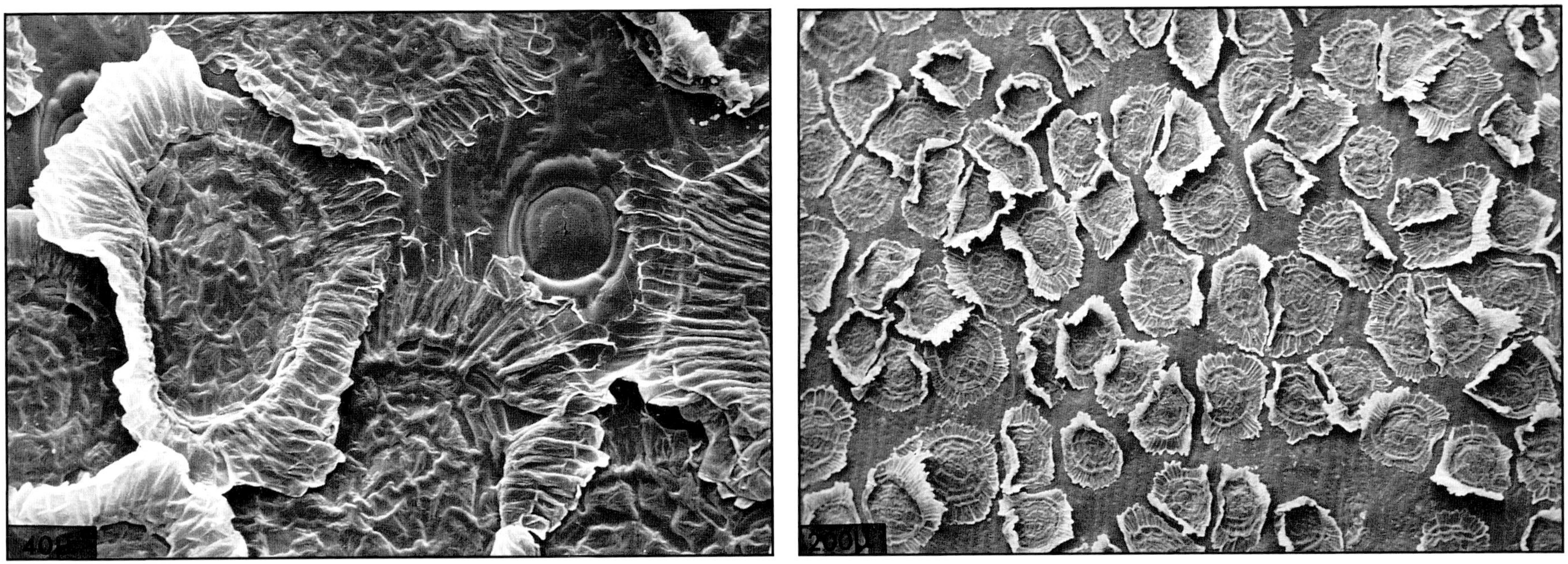

Tillandsia balbisiana

abaxial X400 adaxial X100

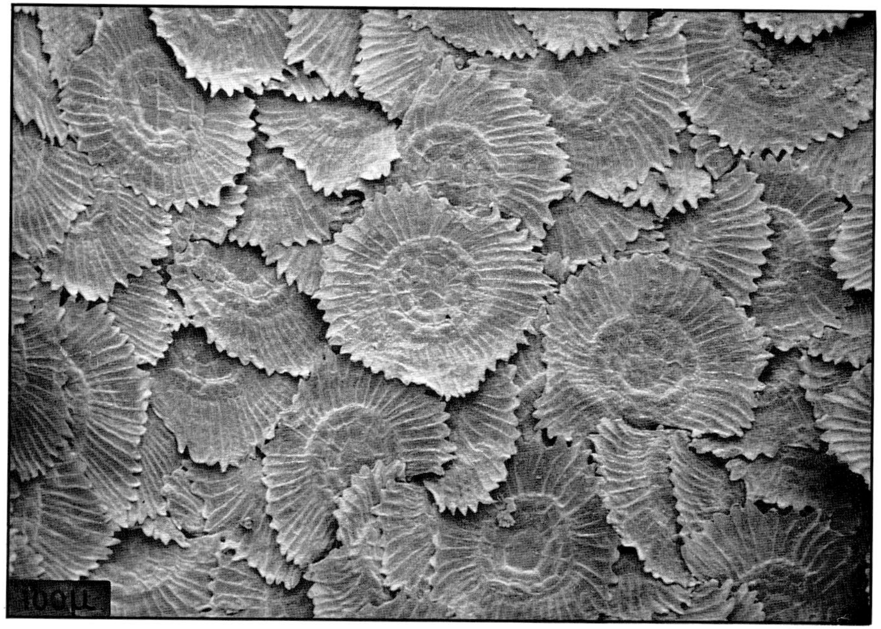

Tillandsia bergeri
abaxial X130 adaxial X180

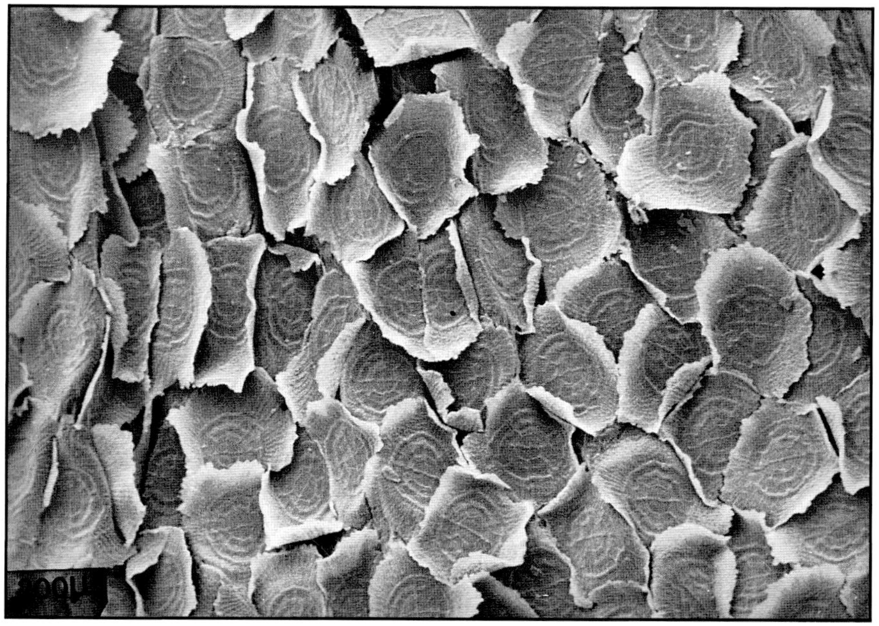

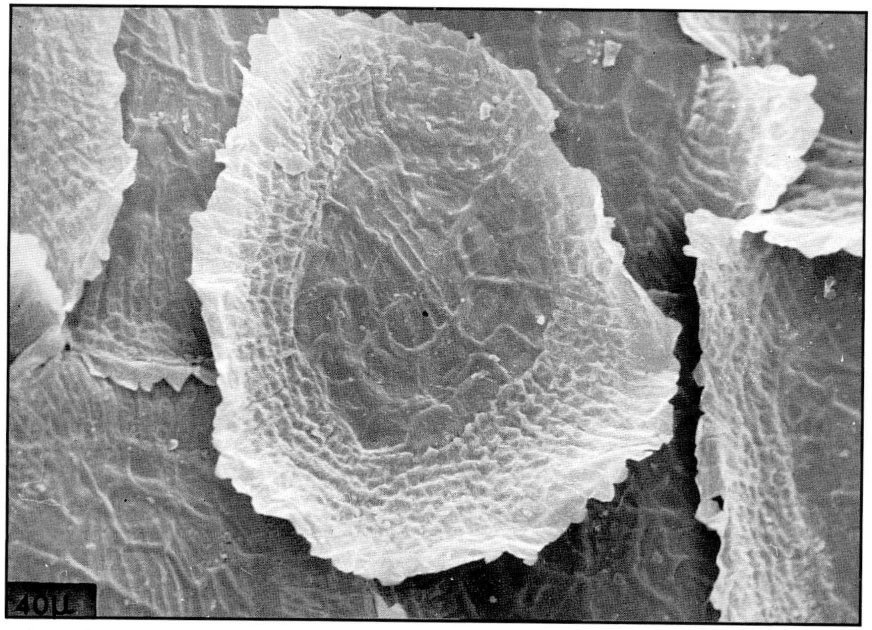

Tillandsia brachycaulos
abaxial X90 adaxial X413

Tillandsia brachycaulos v. *multiflora*
abaxial X375 adaxial X440

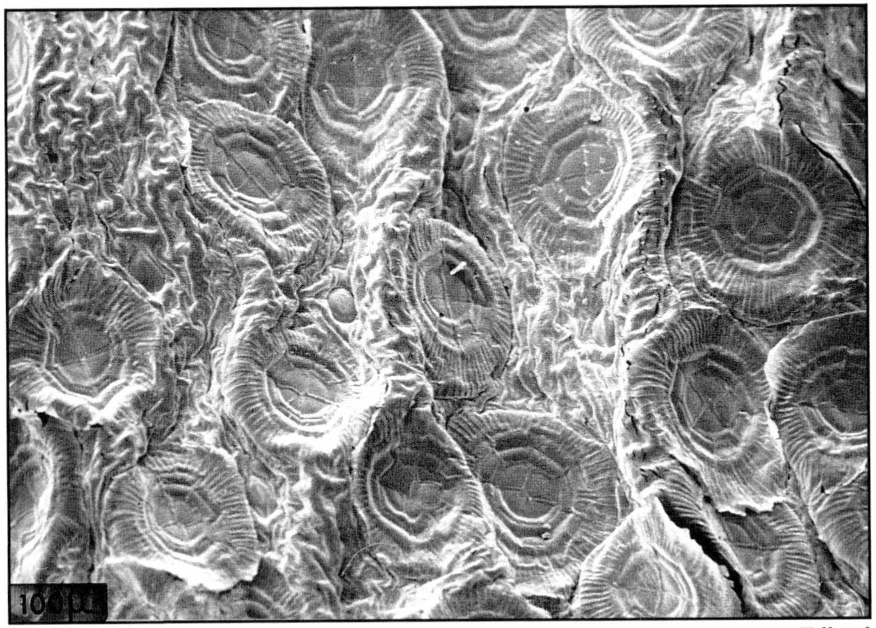

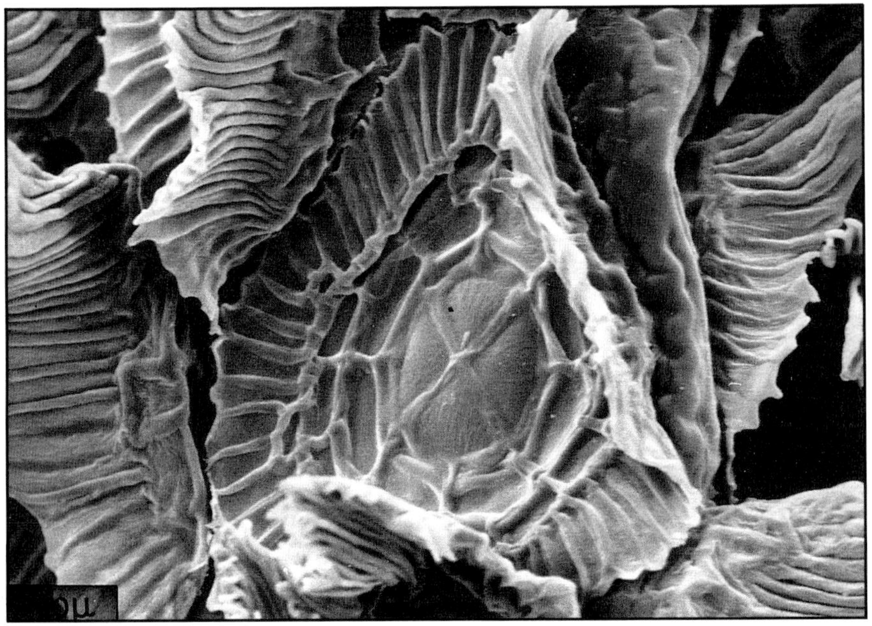

Tillandsia bulbosa
abaxial X180 adaxial X460

Tillandsia butzii
adaxial X190

abaxial X480

Tillandsia caput-medusae
adaxial X200

abaxial X100

Tillandsia cacticola
adaxial X100
abaxial X90

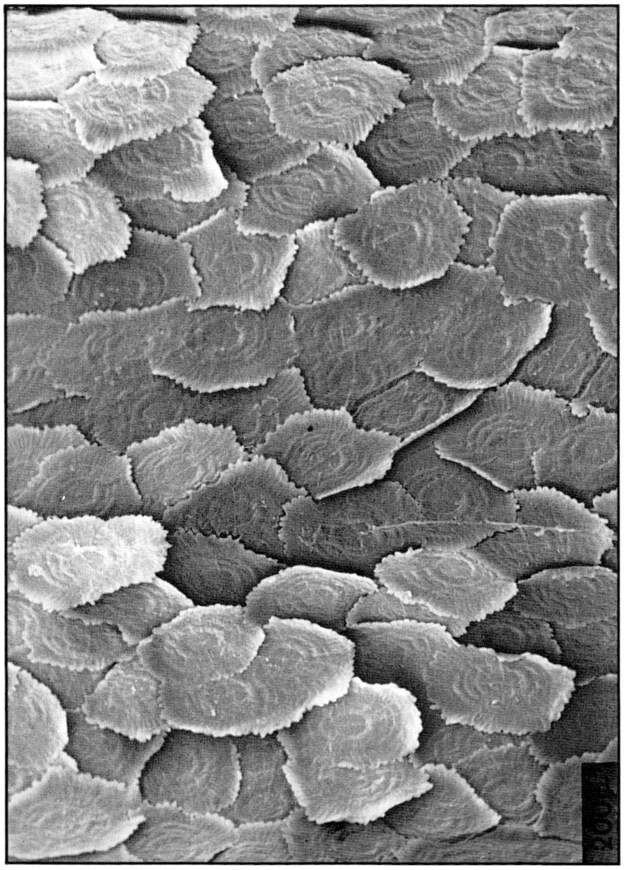

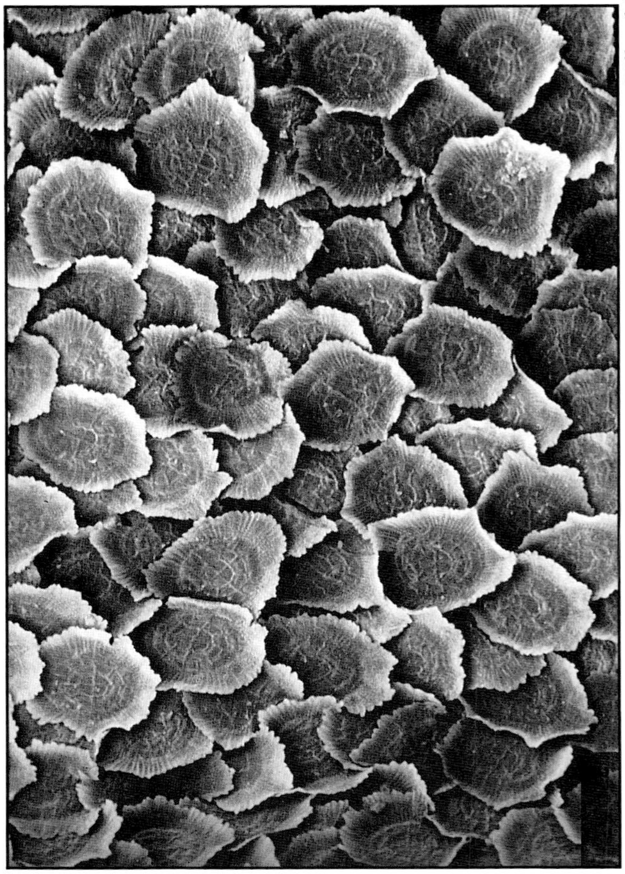

Tillandsia concolor
adaxial X95
abaxial X95

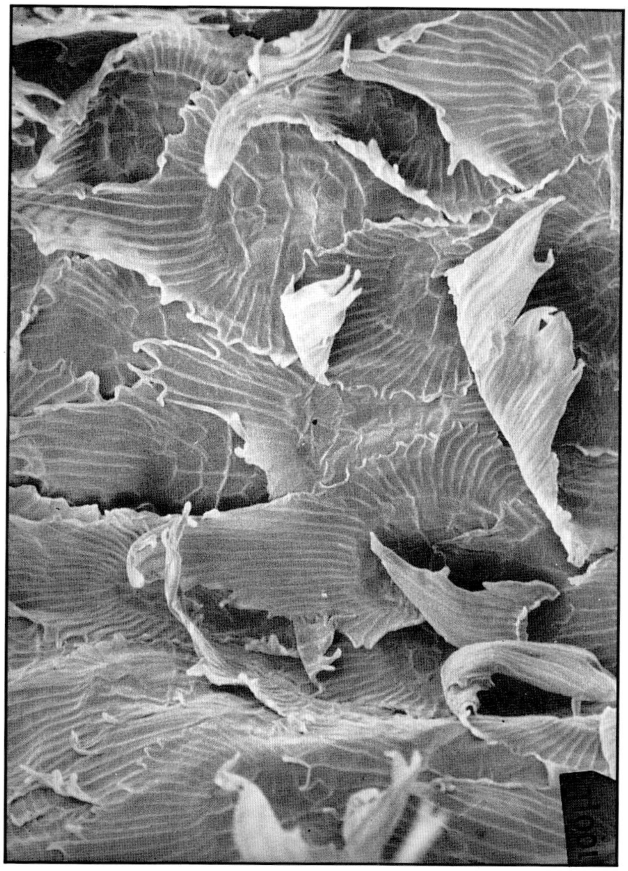

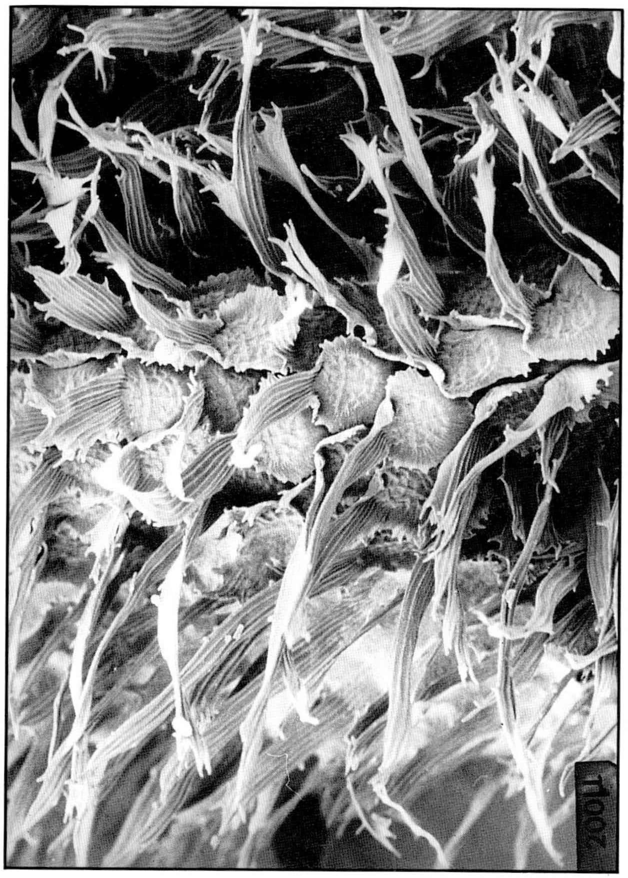

Tillandsia crocata
adaxial X190
abaxial X90

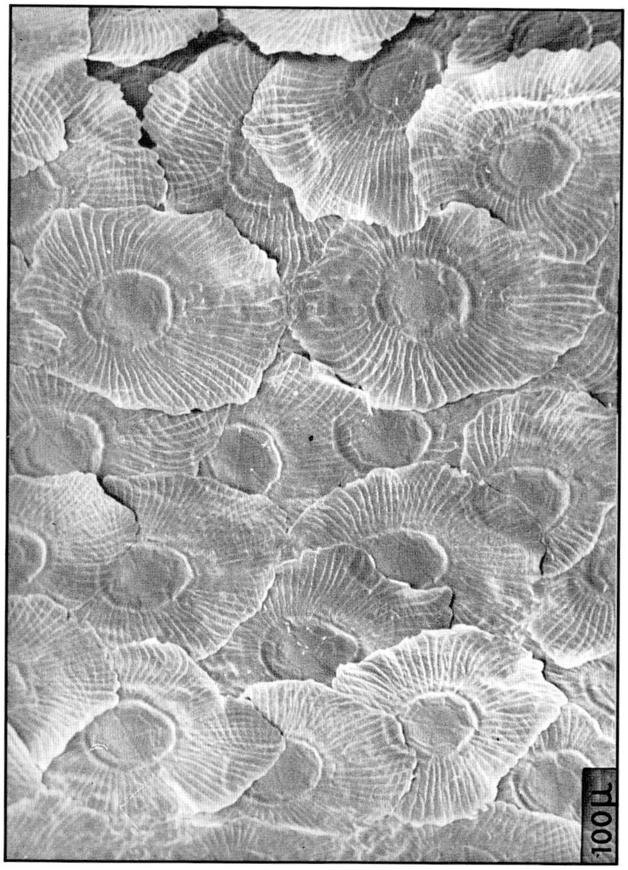

Tillandsia cyanea
adaxial X180
abaxial X190

Tillandsia diaguitensis
adaxial X200
abaxial X200

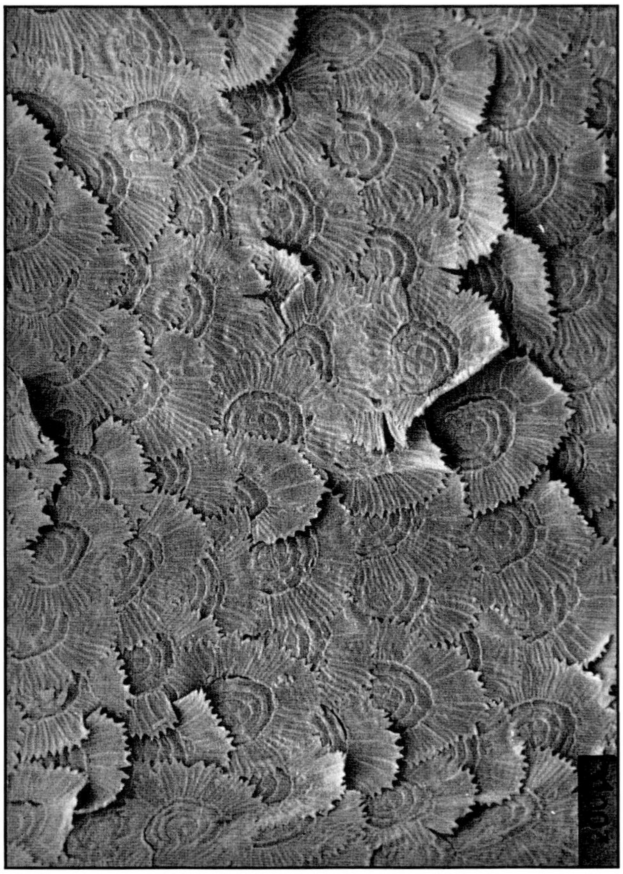

Tillandsia disticha 'Major'
adaxial X190
abaxial X95

Tillandsia duratii
adaxial X330 abaxial X100

Tillandsia edithae
adaxial X88 abaxial X220

Tillandsia fasciculata
abaxial X160 adaxial X460

Tillandsia filifolia
abaxial X340 adaxial X475

Tillandsia friesii

adaxial X160

abaxial X180

Tillandsia ionantha

adaxial X90

abaxial X325

Tillandsia ionantha v. *scaposa*
abaxial X180 adaxial X185

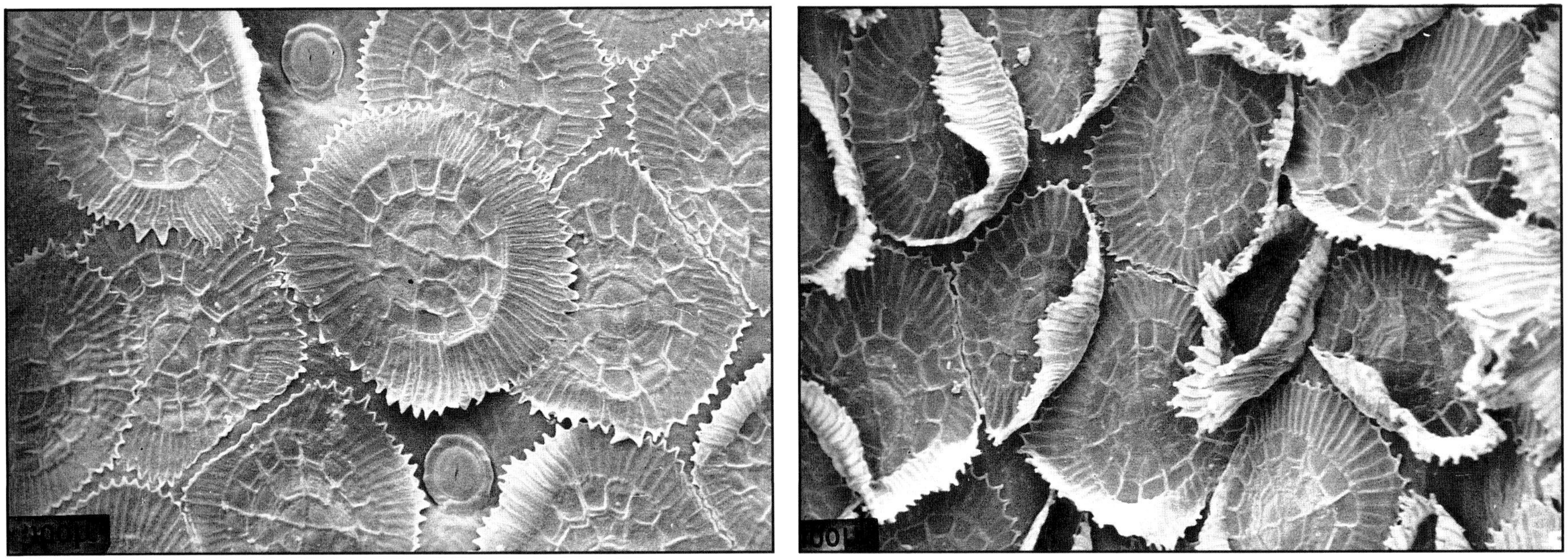

Tillandsia incarnata
abaxial X475 adaxial X108

Tillandsia ixioides
abaxial X190 adaxial X180

Tillandsia juncea
abaxial X325 adaxial X375

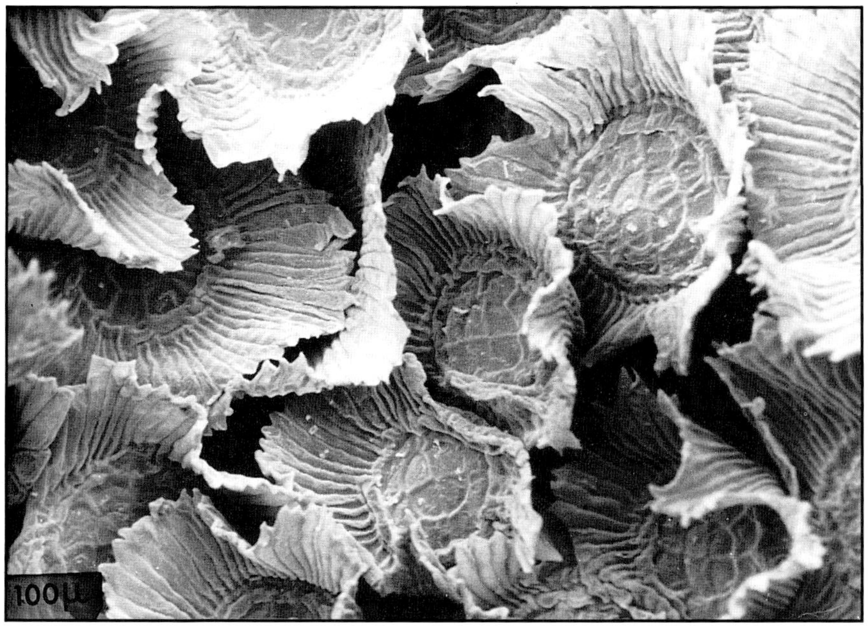

Tillandsia latifolia
abaxial X130 adaxial X200

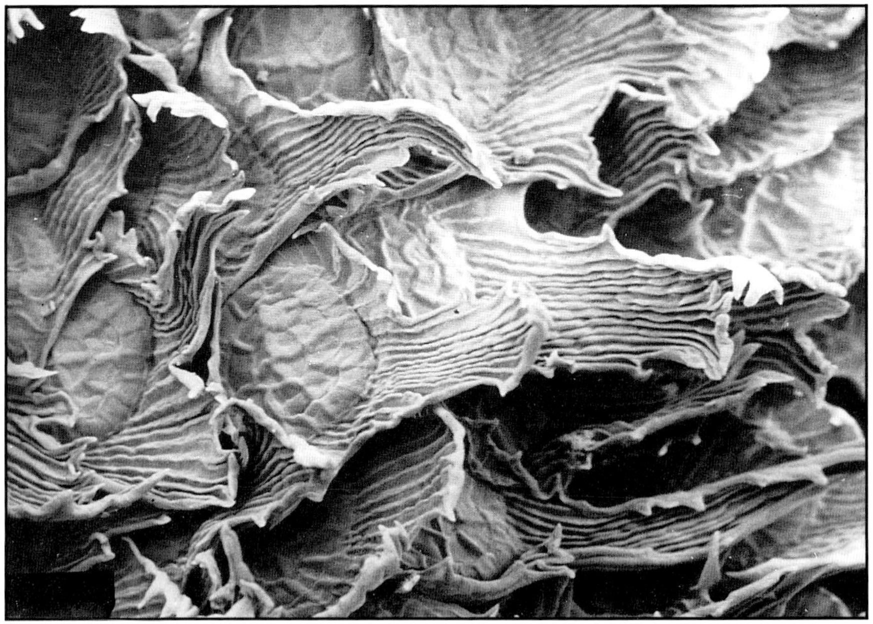

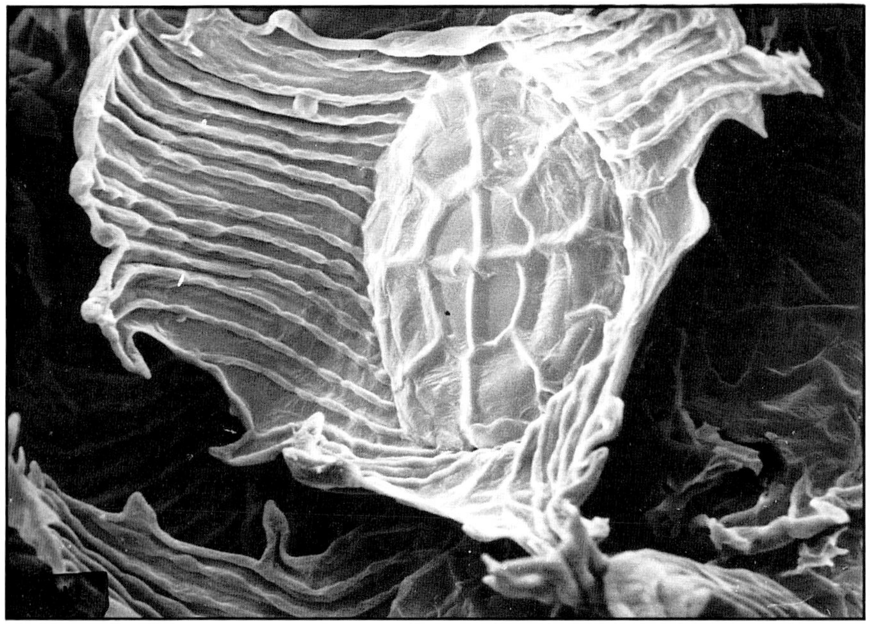

Tillandsia lorentziana
abaxial X190 adaxial X410

Tillandsia meridionalis
adaxial X95
abaxial X200

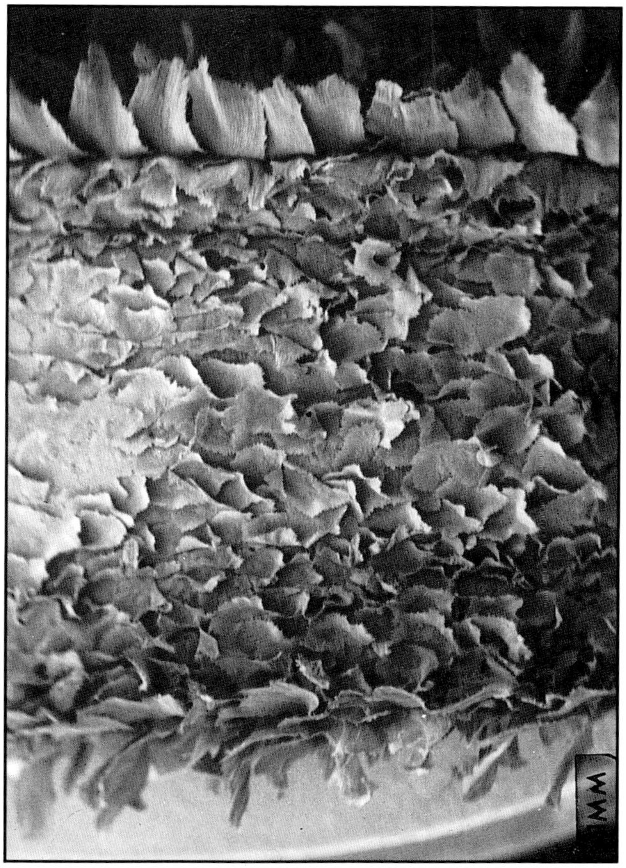

Tillandsia muhrii
adaxial X26
abaxial X190

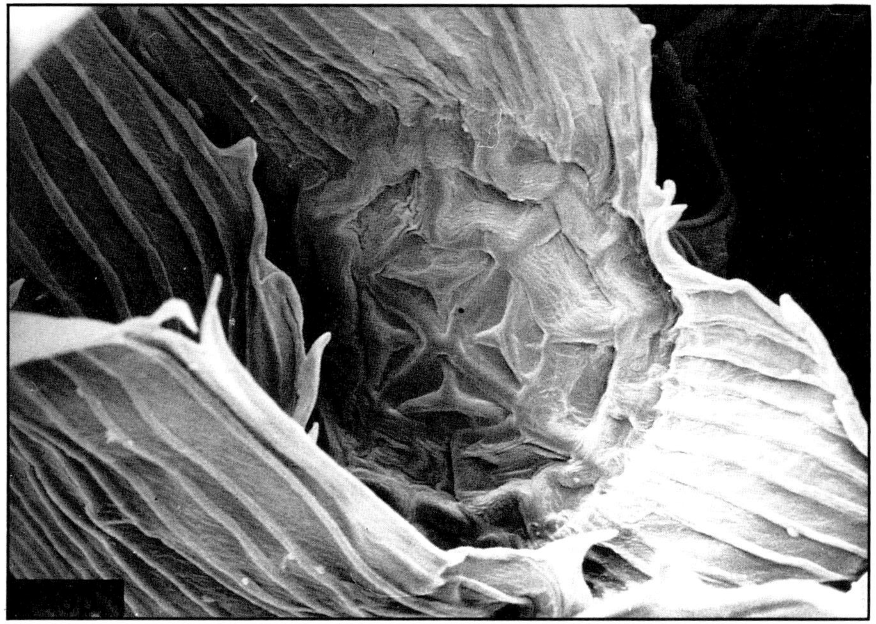

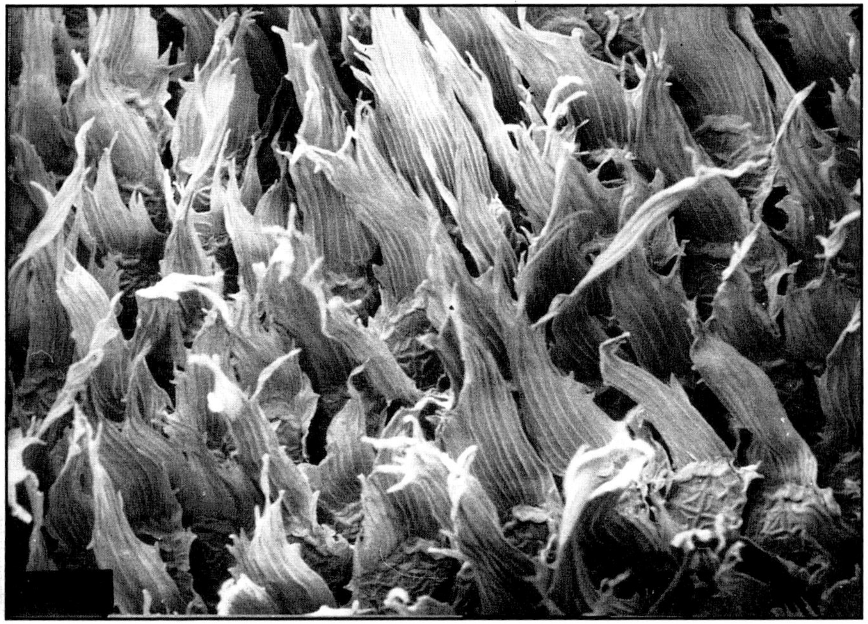

Tillandsia paleacea
abaxial X450 adaxial X100

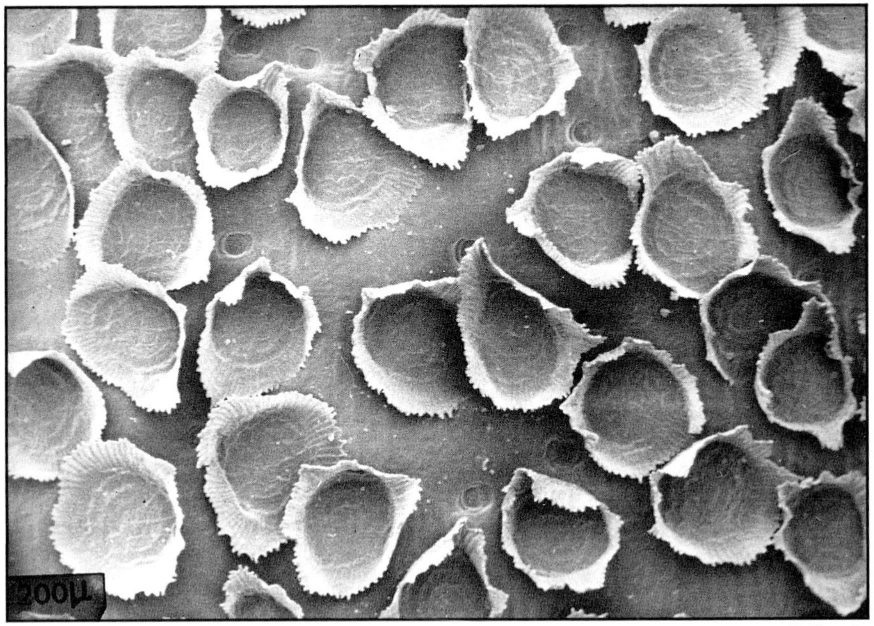

Tillandsia paucifolia
abaxial X95 adaxial X92

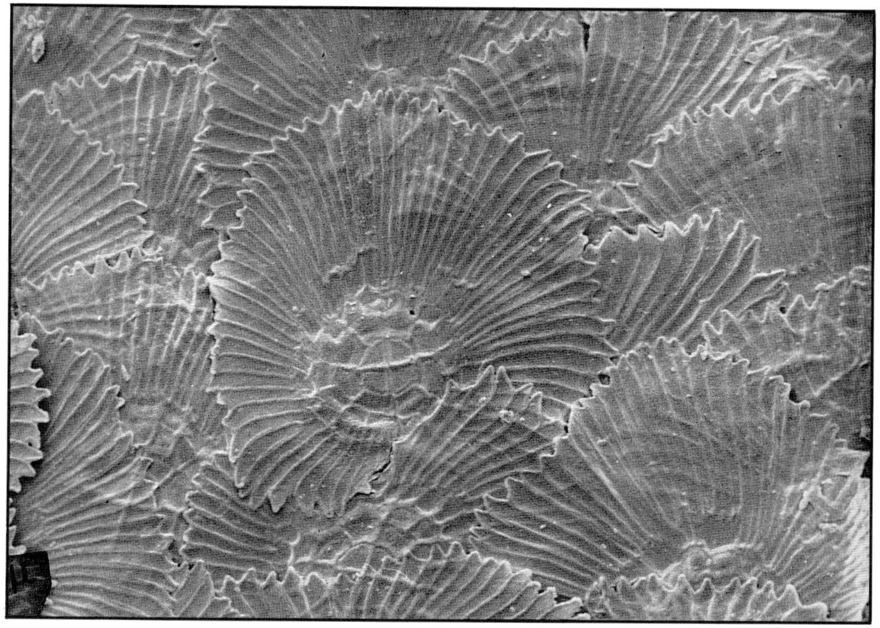

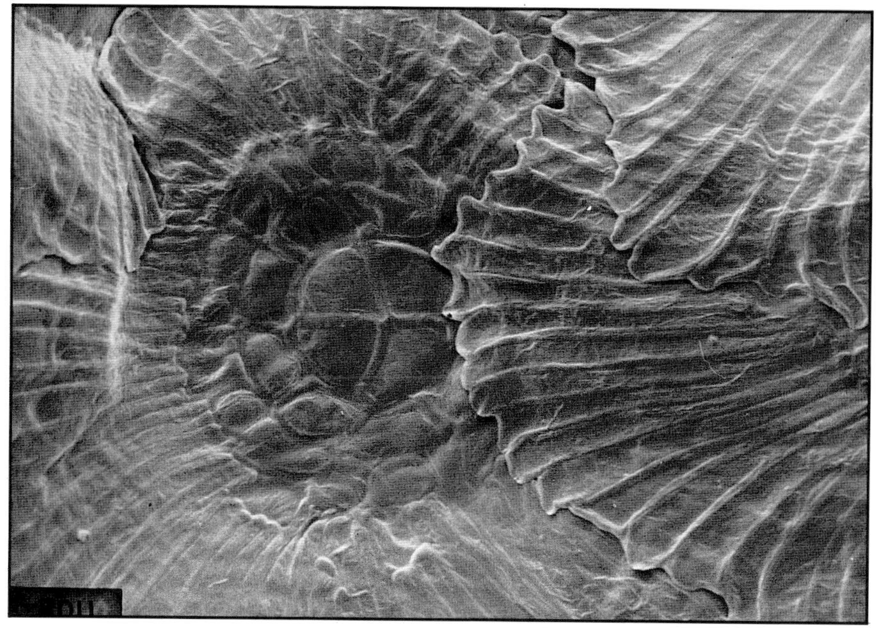

Tillandsia peiranoi

abaxial X170 — adaxial X450

Tillandsia plumosa

abaxial X190 — adaxial X95

Tillandsia polystachia
adaxial X475
abaxial X190

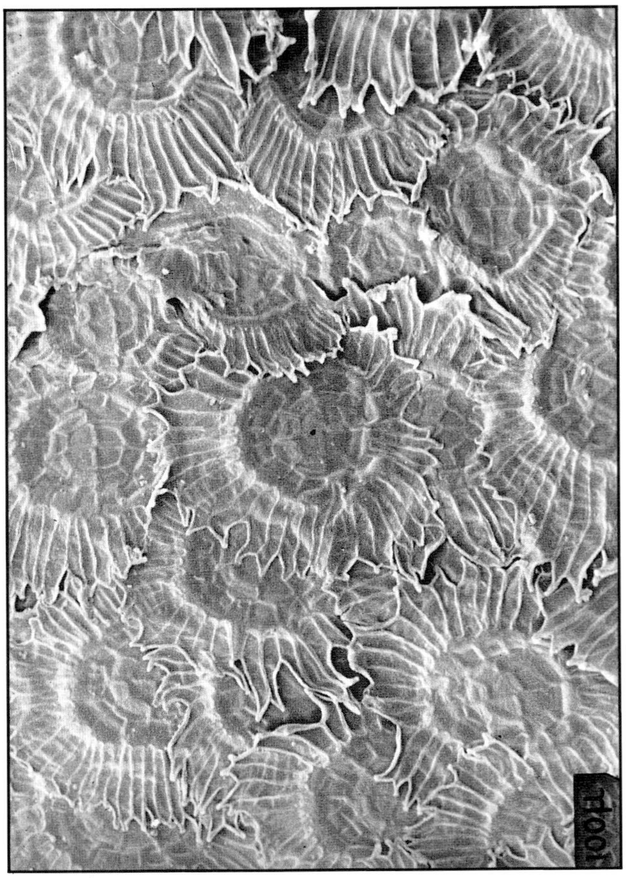

Tillandsia pseudobaileyi
adaxial X190
abaxial X190

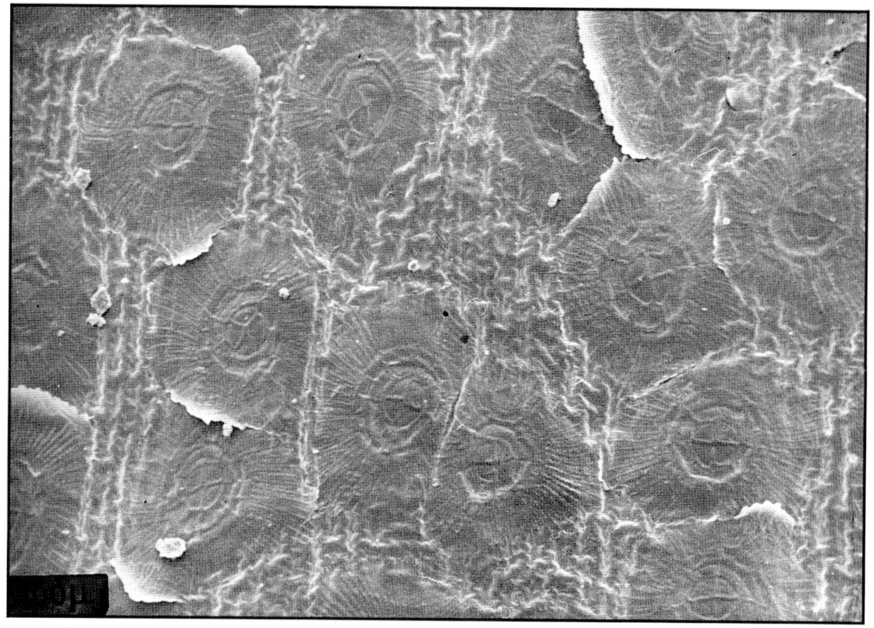

Tillandsia punctulata
abaxial X95 adaxial X180

Tillandsia purpurea
abaxial X200 adaxial X925

Tillandsia reichenbachii
adaxial X475
abaxial X180

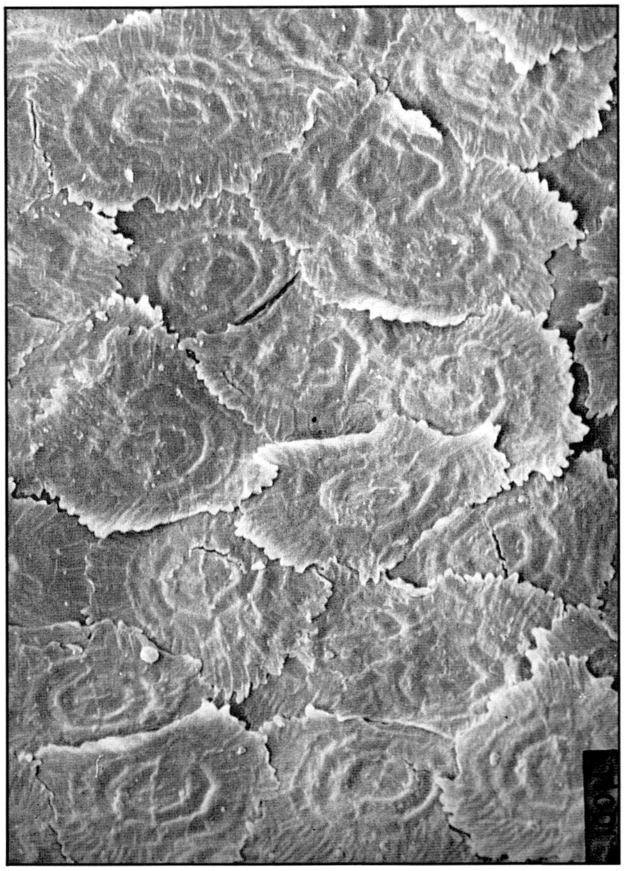

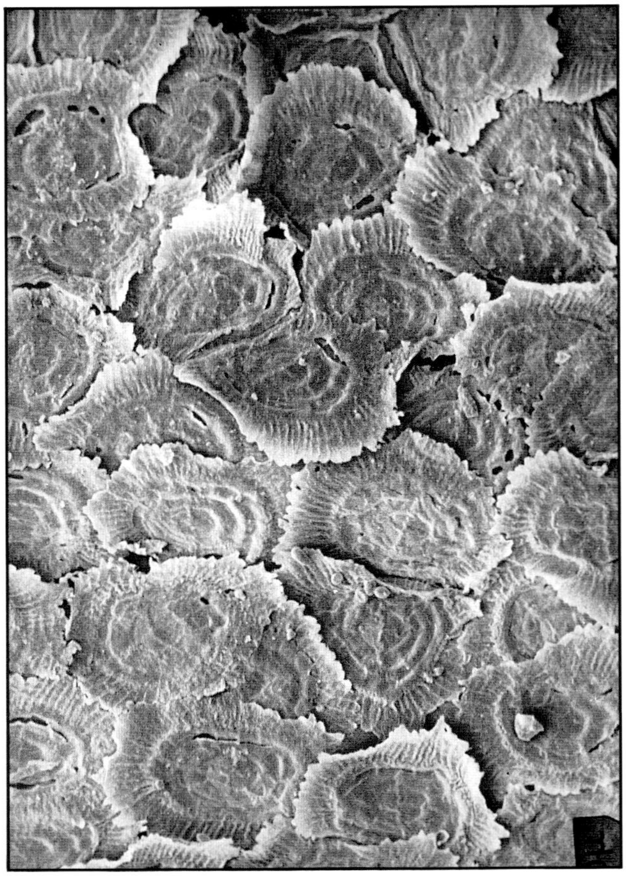

Tillandsia rothii
adaxial X180
abaxial X170

Tillandsia schiedeana 'Major'
abaxial X200 adaxial X220

Tillandsia schiedeana 'Minor'
abaxial X90 adaxial X130

Tillandsia secunda

abaxial X375 adaxial X180

Tillandsia seleriana

abaxial X90 adaxial X180

Tillandsia sphaerocephala
abaxial X240 adaxial X150

Tillandsia straminea
abaxial X190 adaxial X100

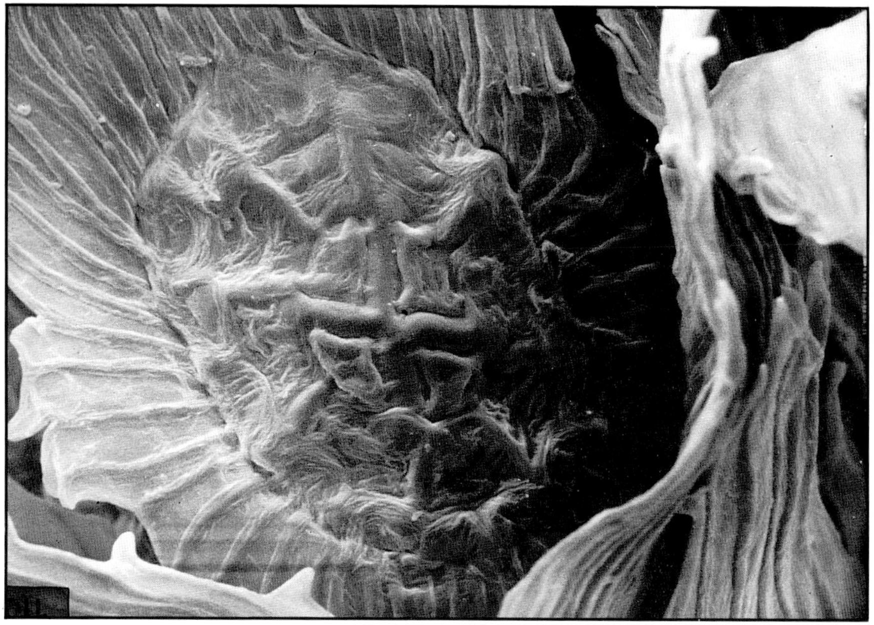

Tillandsia streptocarpa
abaxial X220 adaxial X550

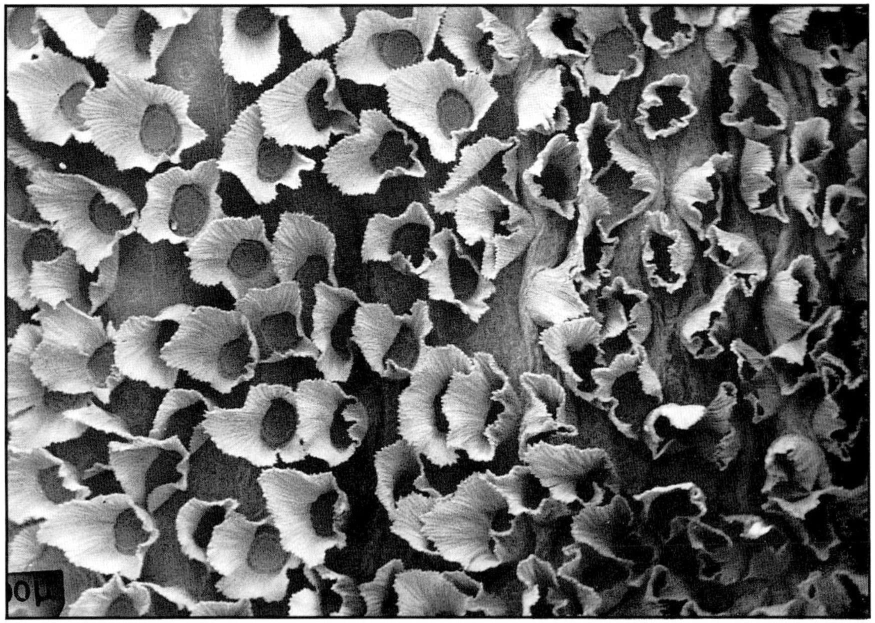

Tillandsia streptophylla
abaxial X46 adaxial X475

Tillandsia stricta
adaxial X95
abaxial X190

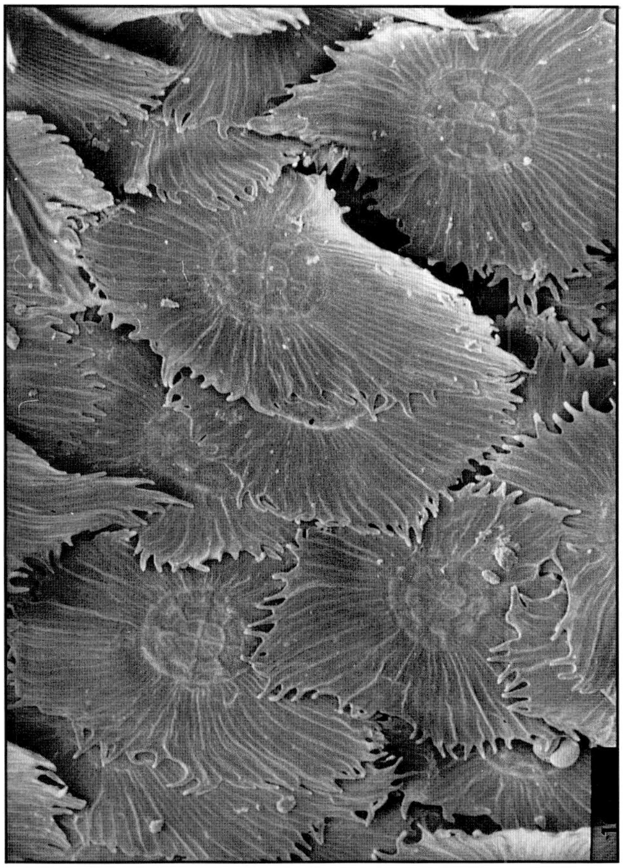

Tillandsia tectorum
adaxial X120
abaxial X190

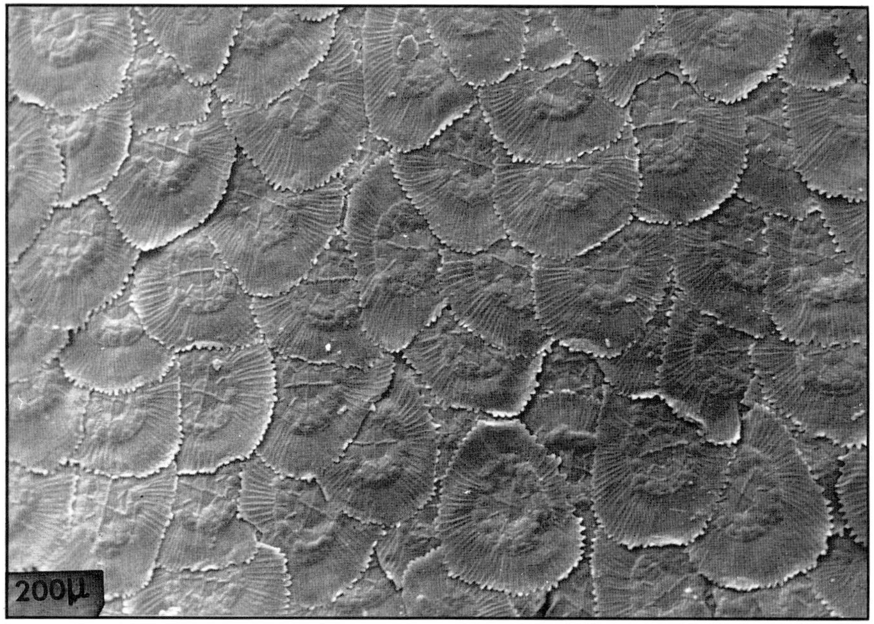

Tillandsia tenuifolia
abaxial X95 adaxial X190

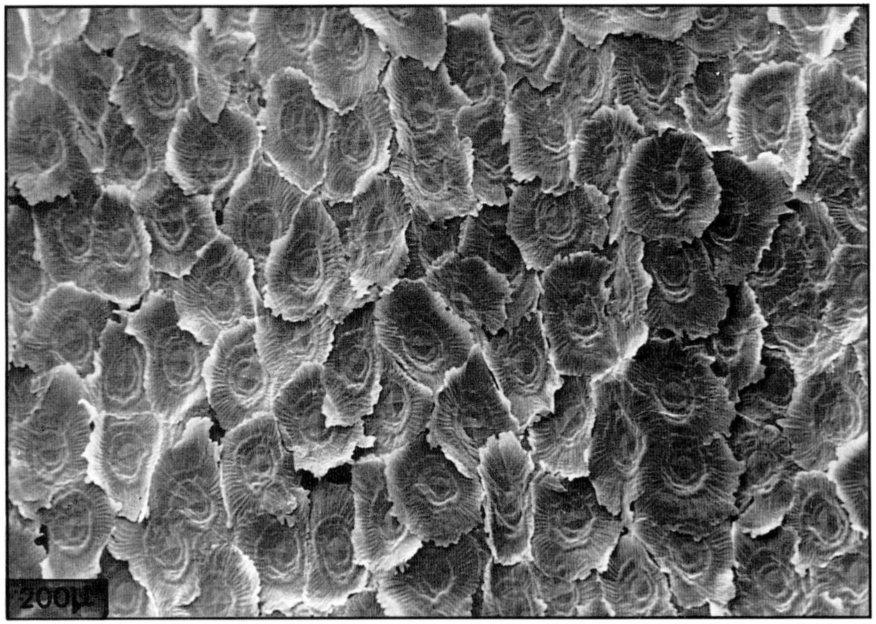

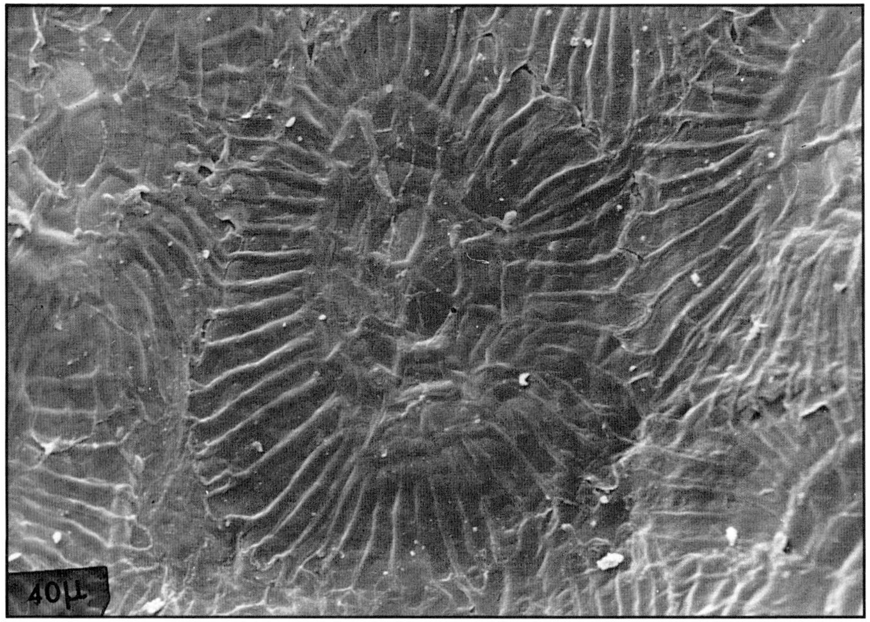

Tillandsia tricolor
abaxial X90 adaxial X475

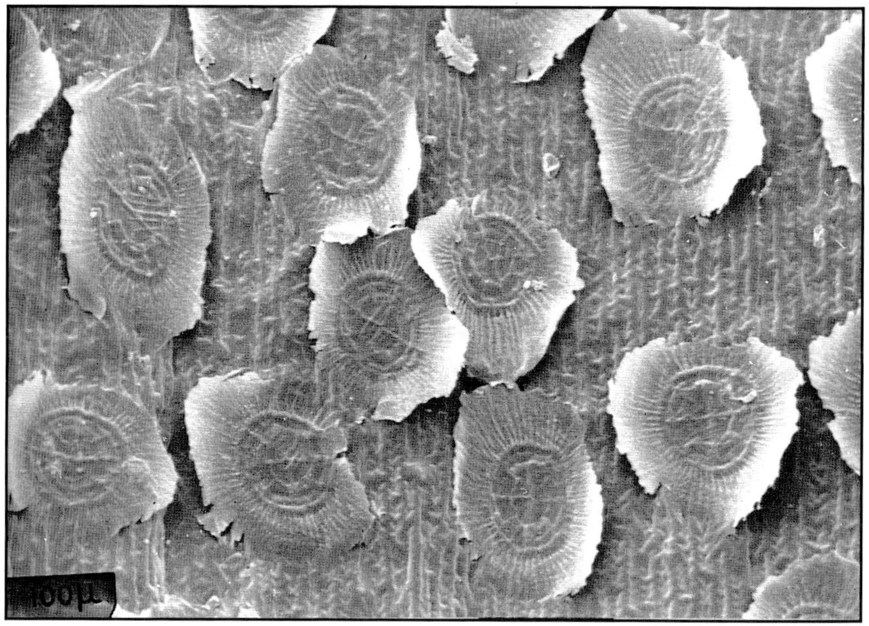

Tillandsia utriculata
abaxial X100 adaxial X180

Tillandsia vicentina
abaxial X200 adaxial X450

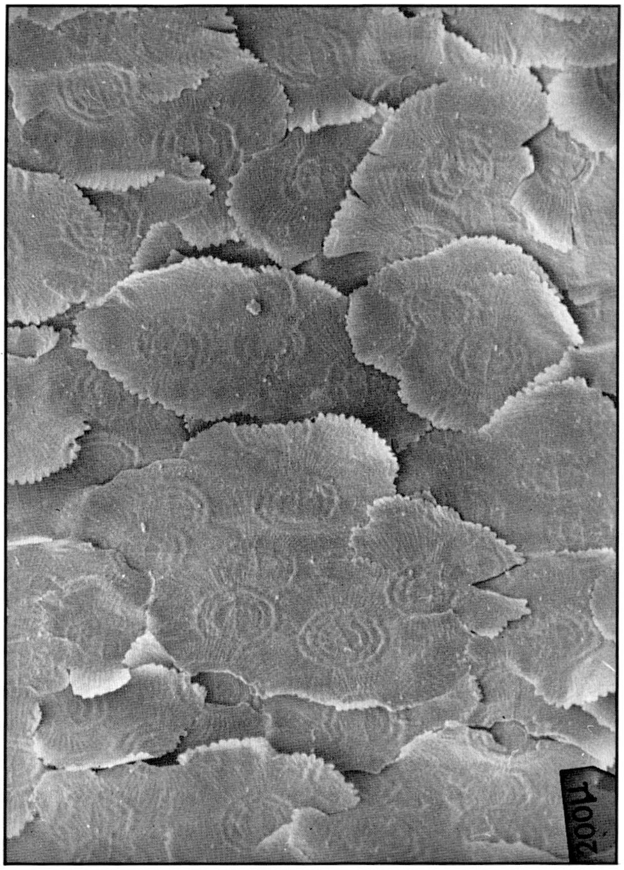

Tillandsia xerographica
adaxial X95
abaxial X388

Tillandsia xiphioides
adaxial X85
abaxial X200

Tillandsia xiphioides v. *tafiensis*
adaxial X190
abaxial X93

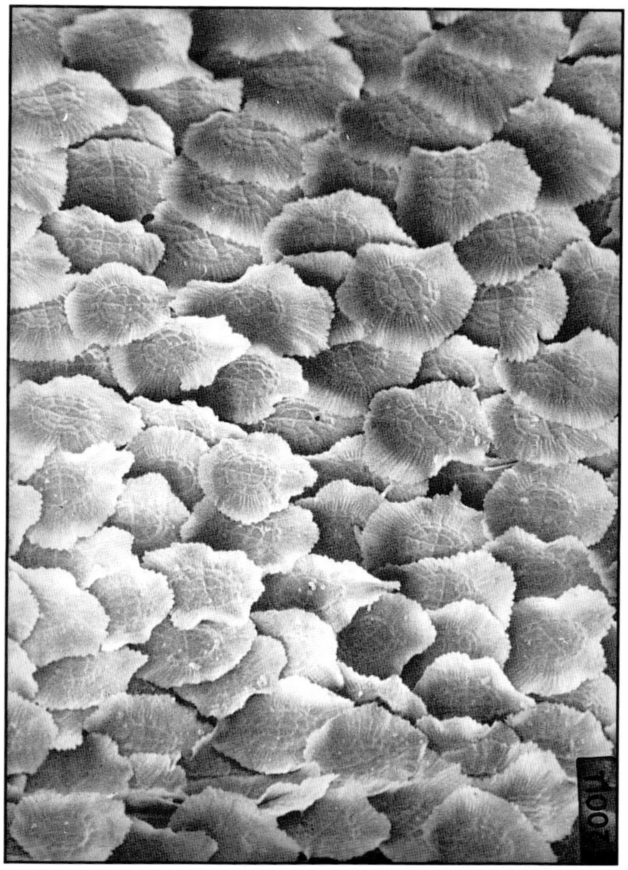

Vriesea espinosae
adaxial X100
abaxial X375

Color Chart

	CIRCLE OF SPECTRUM COLORS	GRADATION OF DARKNESS TONES: TINTS *achieved by adding % of light or white*			GRADATION OF DARKNESS TONES	GRADATION OF DARKNESS TONES: SHADES *achieved by adding % of shadow or black*		
		1 PALE	2 LIGHT	3 MODERATE	4 PURE COLOR	5 STRONG	6 DARK	7 BLACKISH
XANTHIC COLORS	YELLOW 360°	1 CREAM	2 MIMOSA	3 CANARY	4 BUTTERCUP YELLOW	5 OLD GOLD	6 LIGHT BROWN	7 CHOCOLATE
	YELLOW ORANGE 30°	8 IVORY	9 BUFF	10 CHROME YELLOW	11 APRICOT	12 BURNT ORANGE	13 CINNAMON	14 BROWN
	ORANGE 60°	15 PEACH PINK	16 SALMON	17 CARROT RED	18 ORANGE	19 COPPER	20 BRICK RED	21 RUST
	ORANGE RED 90°	22 FLESH	23 CORAL	24 MANDARIN RED	25 SCARLET	26 CHERRY	27 BLOOD RED	28 CARDINAL RED
CYANIC COLORS	RED 120°	29 SHELL PINK	30 CLEAR PINK	31 CARMINE	32 CRIMSON	33 RUBY RED	34 WINE RED	35 GARNET LAKE
	RED PURPLE 150°	36 BLUSH PINK	37 ROSE PINK	38 ROSE	39 PURPLE	40 AMARANTH	41 BEET PURPLE	42 MAROON
	PURPLE 180°	43 LAVENDER	44 LILAC	45 MAUVE	46 AMETHYST	47 ROYAL PURPLE	48 INDIGO	49 PLUM PURPLE
	PURPLE BLUE 210°	50 ORCHID	51 WISTERIA BLUE	52 ASTER VIOLET	53 PURPLISH BLUE	54 VIOLET	55 MOORISH BLUE	56 GRAPE VIOLET
	BLUE 240°	57 SKY BLUE	58 FLAX BLUE	59 FRENCH BLUE	60 ULTRAMARINE	61 CORNFLOWER BLUE	62 PRINCESS BLUE	63 NAVY BLUE
	BLUE GREEN 270°	64 AQUAMARINE	65 TURQUOISE	66 MING GREEN	67 JADE GREEN	68 PINE GREEN	69 BOTTLE GREEN	70 IVY GREEN
FOLIAR COLORS	GREEN 300°	71 SEAFOAM	72 SPRING GREEN	73 GRASS GREEN	74 PARIS GREEN	75 EMERALD GREEN	76 MEADOW GREEN	77 FOREST GREEN
	YELLOW GREEN 330°	78 CHARTREUSE	79 CITRON	80 NILE GREEN	81 LETTUCE GREEN	82 MOSS GREEN	83 FERN GREEN	84 OLIVE GREEN

Reprinted from *Exotica* with permission of Dr. A.B. Graf (see Acknowledgments).

Glossary

a-. A prefix denoting "without" or "not."

abaxial. Situated facing away from the axis of the plant, i.e. the under or outer surface of a leaf. *Syn.* **dorsal**. *Ant.* **ventral**.

-aceae. Latin suffix added to stem of name or synonym of type for the genus in names of plant families. Means "resembling" or "similar to."

acid. Substance that has a pH below 7.0; disassociates and releases hydrogen (H+) ions. *Compare* **base.**

acrotonal. Offsets emanating from the apex of a caulescent plant, i.e. from the base of the inflorescence.

active transport. An energy requiring process by which a cell moves a substance (e.g. nutrients) across the plasma membrance, often against a concentration gradient, replacing what is used in food manufacturing or other respiratory activities.

ad-. A prefix meaning "to" or "toward."

adaxial. The side toward the axis; the surface of a leaf that faces the stem, i.e. the upper or inner side of a leaf. *Syn.* **ventral.** *Ant.* **dorsal.**

adenosine triphosphate (ATP). An organic molecule containing high-energy phospate bonds, the major source of usable energy in metabolism. Most frequently found in mitochondria.

adnate. The union of unlike parts, e.g. stamen filaments to petals or an ovary to a calyx tube.

adventitous (pup). Plant organs produced in an unusual or irregular position or at an unusual time of development, as in adventitious shoots.

Aechmea. Bromeliad genus described by Ruiz & Pavon in 1794 after the Greek *aechmea*, or "spear tip"—a reference to the points on the perianth.

ageotropism. The absence of any tendency of an organ or organism to grow in a given direction because of the force of gravity.

air fern. The dead skeleton of a marine animal of the genus *Bubula*. Because of its fernlike growth when dried, it is sold as a novelty—a "living fern," after having been dyed green.

alkaline. Above pH 7.0; a base.

allele. One of the two or more alternative forms of a gene that occurs at the same locus on homologous chromosomes; because alternate forms exist, an organism may be homozygous for some alleles and heterozygous for others, and would then be subject to the rules of Mendelian inheritance.

allopatric, allopatry. Species of populations originating in or occurring in different geographical regions, without overlapping distributions. *Compare* **sympatric.**

ambient. Relating to the environment at a particular time, as in ambient temperature. Prevailing.

amino acids. The primary components of proteins, each containing an amino group and linked together by peptide bonds; large numbers of amino acids connected together in this manner form proteins.

an-. A prefix meaning "without" or "not." Equivalent to the prefix **a-**.

analogy. The concept that certain organs may have a similar function and appearance, yet are not related in terms of embryonic or phylogenetic origin. *Contrast* **homology.**

androgynous. Having staminate and pistillate flowers in the same inflorescence.

angiosperm. Any flowering plant; any plant of the Angiospermae; a plant bearing seeds that develop in an enclosed ovary or carpel.

anterior. 1. On the front side. 2. In the flower, the side away from the axis and toward the subtending bract.

anther. The pollen-forming part of the stamen.

anthesis. The time at which a flower comes into full bloom; when a flower is fully expanded.

anthocyanin. Natural pigments of blue, purple, or red that are especially common in the petals of flowers; these pigments vary in color according to the acidity of the cell sap.

apex, *pl.* **apices.** At, near, or belonging to the tip or point.

apical meristem. A group of meristematic cells at the tip of a root or shoot from which all the tissues of the mature axis are ultimately formed; these may be vegetative or reroductive cells.

appressed. Closely and flatly pressed against the entire length of an organ or part; with an angle of divergence of 15 degrees or less.

arroyo. A dry gully or watercourse.

asexual. Without sex, as in vegetative propagation.

atmospherics. A word first used by Mez to describe the xeric tillandsioids that have neither absorbing roots or tank impoundments. These plants must secure moisture and nutrients directly from rainfall and canopy fluids by way of their absorbing foliar trichomes.

attenuate. Narrowing to a point; thin or slender.

autotroph. An organism that makes its own food.

awl-shaped. Narrow and gradually tapering to a sharp point; sharp-pointed from a broader base.

axil. The angle between the stem axis and a leaf petiole, branch or other appendage attached to it.

axis, *pl.* **axes.** The main stem of a plant; the main or central line of development of any plant or organ.

backcross. Cross fertilization of a hybrid with one of its parents.

BARRY, DAVID (1898-1978). Was the first to conceive the idea of an international bromeliad society. With his unflagging enthusiasm and large supply of commercial plants, he was a driving force in the popularization of bromeliads in United States.

base. The lower or bottom portion of a plant organ. A substance that has a pH above 7.0; a compound that reacts with an acid to form a salt and has a tendency to acquire a hydrogen atom. *Compare* **acid.**

basitonal. Offsets that grow from the base of the plant.

BENZING, DAVID (1938-). Research scientist, botanist, author, and biology professor at Oberlin College in Ohio. A leading authority on the biology of bromeliads.

Billbergia. Bromeliad genus described in 1821 by Thunberg for Gustave Billberg, a Swedish botanist.

binomial. Scientific name of an organism consisting of two Latin names; the genus name is capitalized followed by the lower case species. Names always *italicized* or underlined.

bipinnate. Twice pinnate; a once-compound inflorescence in which the flowers are borne on secondary spikes rather than on the simple extension of the scape.

blade. Upper portion of leaf or petal. *Compare* **sheath.**

bordered pit. Pits in which the secondary wall arches over the pit membrane; typical of gymnosperm tracheids.

bract. A modified, often much reduced leaf subtending a flower or inflorescence.

bromeliad. Common name for a family of herbs confined to the Western Hemisphere, with the exception of *Pitcairnia feliciana*, which occurs on the west coast of Africa.

BROMELIUS, OLAUS OLAI (1639-1705). Swedish botanist after whom the genus *Bromelia*, and later the family Bromeliaceae, was named.

BROWN, GREGORY (1951-). Botanist, taxonomist, and professor at the University of Wyoming who specializes in bromeliad chromosome research and taxonomic relationships.

Calvin-Benson (C3) pathway. The carbon fixation pathway common to all green higher plants.

calyx, *pl.* **calyces**. A collective term for all the sepals of a flower.

canalizing. Process favoring a particular morphological developmental pattern. Once initiated, it results in the achievement of a standard phenotype.

canescent. Covered with dense, fine, grayish-white hairs; becoming hoary, usually with a gray pubescence.

capitate. Formed like a head; aggregated into very dense clusters or heads, especially in reference to densely congested floral spikes.

carpel. The ovule-bearing structure of a flower.

castaneous. Reddish brown; chestnut-colored; dark brown. Often applied to a leaf sheath.

cataphyllic. Small, scaly leaves covering stolons that often die when mature, but remain as a tough indumentum.

Catopsis. Epiphytic bromeliad genus described by Grisebach in 1864. The word is from the Greek and means "view." *Catopsis* are often identified by their waxy, glaucous, green leaves.

caulescent. Having an obvious leafy stem.

cavitation. The formation of gas bubbles in water, as in xylem tissue.

cellulose. A complex carbohydrate, the chief component of the cell walls; it consists of long chainlike molecules of glucose that form microfibrils.

central disc cells. The four concentric cells making up the center of the trichome shield.

channeled. Deeply grooved lengthwise.

chartaceous. Papery in texture, opaque and thin.

chlorophyll. The green pigments found in the thylakoids of chloroplasts, which are essential for the utilization of light energy in photosynthesis.

chloroplast. A subcellular organelle that contains the photosynthetic, membrane-bound apparatus of green plants; it contains chlorophyll.

chromosome. The self-duplicating structural units of the cell nucleus containing genes in a linear order. They are threads of chromatin composed of nucleic acids and proteins. Chromosomes are normally dispersed between cell divisions, becoming condensed and microscopically visible only during mitosis and meiosis.

ciliate. Fringed with conspicuous hairs along the margin.

cinereous. Ash-colored; light gray.

claw. *See* **petal claw**.

clone. A population of individuals descended from a single parent by asexual reproduction.

cohesion ratio. Term coined by Lyman B. Smith in 1934 to describe the amount of geographical overlap of all species in a subgenus relative to the total distribution of those species.

coma. A tuft of appendages, such as the collection of hairs borne on the seeds of members of subfamily Tillandsioideae.

companion cell. A small parenchyma cell intimately associated with a sieve tube element and probably aiding the latter in its food transporting function.

compensation point. The level of light intensity at which oxygen production during photosynthesis is equal to oxygen consumption by respiration.

complanate. Flattened; compressed.

compound inflorescence. Inflorescence composed of secondary branches.

con-. A prefix meaning "with" or "together."

congested. Crowded. *Compare* **lax**.

conglutinated. United firmly together; glued together.

connate. Union or fusion of like parts or organs to one another.

contorted. Twisted around a central axis; bent or distorted.

convergent evolution. The independent development of similar structures among forms of life that are only distantly related and may be separated by great geographic distances, e.g the Euphorbiaceae in Africa and the Cacteceae in the New World.

coriaceous. Thick, tough, and leathery.

cornicle. In insects, terminal, dorsal, erect or semierect tubules that secrete a waxy, defensive liquid.

corolla. A collective term for the petals of a flower.

Cryptanthus. Small, terrestrial bromeliad genus from Brazil described in 1836 by Otto and Dietrich. Common name is "earth stars" because of flat leaf form and symmetrical markings. *Crypt* is from the Latin for "hidden" and *anthus* is from the Greek for "flower."

cryptic. Self-concealing.

cryptogam. A plant that reproduces by spores instead of seeds, e.g. ferns, mosses, and algae. *Compare* **phanerogam**.

cultivar. A plant type within a cultivated species that has recognizably different characteristics. A contraction of "cultivated variety." Also used to distinguish a natural variant in a wild population of plants that has not yet been described in Latin.

-cyte, cyto-. A suffix or prefix meaning "pertaining to a cell."

cytoplasm. The matrix of a cell in which the organelles are located.

decurved. Curved downward; bent downward.

deflexed. Turned downward abruptly.

dehiscence. To split open when ripe, usually along definite lines or sutures in order to release seeds.

desiccated. Dry; without moisture.

determinate. Organ or organism with a preprogrammed, limited growth duration.

detritus. Any particulate accumulation of disintegrated animal, mineral, or vegetable matter.

Deuterchonia. Bromeliad genus described by Mez in 1894 for Ferdinand Julius Cohn. Comprised of seven species. Native to South America.

dicotyledon. A member of the Dicotyledoneae; one of the two taxa of angiosperms (class Magnoliopsida) usually characterized by the following: two seed leaves (cotyledons), flower parts in fours or fives (or multiples of these), leaves with net venation, and root systems with tap roots.

differentially (semi) permeable membrane. A membrane that is permeable to small molecules such as water, but not to certain larger molecules or ions.

diffusion. The movement of suspended or dissolved particles from a more concentrated to a less concentrated region of the diffusing material as a result of the random movement of individual molecules; the process of random movement tends to distribute molecules uniformly throughout a medium.

digitate. Fingerlike; shaped like an open hand; palmate.

dioecious. 1. Having staminate and pistillate flowers on different plants of the same species. 2. Having male and female organs on different plants, e.g. plants in bromeliad genus *Catopsis*.

diploid. A nucleus, organism, or generation that has two sets of chromosomes and thus has a 2n chromosome number. *Compare* **haploid**.

directional selection. A method of natural selection that occurs when the environment is changing progressively in a particular direction, i.e. drier/wetter; hotter/colder.

discrete. Separate, not coalesced.

distichous. Tworanked, with leaves or flowers in two opposite rows in the same plane.

diurnal. Opening or occurring during the daytime.

dorsal. The lower or underside of a leaf. *Syn*. **abaxial**. *Ant*. **ventral**.

Dyckia. Terrestrial bromeliad genus described by Schultes in 1830 for Prince von Salm-Dyck. Comprised of over 100 species. Native to South America.

e-. A prefix meaning "without," "out of," or "from."

ebracteate. Without bracts.

ecto-. A prefix meaning "outside," "out," or "outer."

elliptic. Shaped like an ellipse; broadest at the center, tapering equally to the base and apex.

endemic. Native or confined naturally to a particular and usually very restricted geographical area or region. *Compare* **indigen**.

endo-. A prefix meaning "within" or "inner."

endodermis. A single layer of cells that usually forms a sheath or covering around the vascular region in roots.

entire. Without indentations or incisions on the margin; smooth.

environment. The sum total of the external conditions that affect the growth and development of an organism.

enzyme. A complex protein catalyst produced in living cells that, even in very low concentraton, speeds up the rate of certain chemical reactions but is not consumed in the reaction.

epi-. A prefix meaning "upon," "over," or "beside."

epidermis. The outer layer of cells of the primary tissues of a plant, e.g. of leaves, young stems, and young roots.

epigynous. Having a floral organization in which the ovary occupies an inferior position to the sepals, petals, and stamens. *Compare* **hypogynous** and **perigynous**.

epiphyte. A plant that grows on another plant but does not feed on it.

epithet. A single descriptive word applied to species; makes up the second component of the scientific name.

erect. Applied to flowers or floral bracts that stand up nearly parallel to the rhachis to which they are attached.

erubescent. Blushing rosyred.

eutrope. A plant that grows in a nutrient rich environment. *Compare* **oligotrope**.

ex-. A prefix meaning "out," "off," "from," or "beyond."

exine. The outer wall layer of spores and pollen grains.

exo-. A prefix meaning "out," "outside," or "without."

exserted. Rising above, projecting beyond, or protruding.

extant. Presently living or existing.

extra. A prefix meaning "outside of" or "beyond."

exude. To discharge slowly through small pores.

F_1. The first filial generation (hybrid) in a cross between two parental types.

F_2. The second filial generation resulting from the self-fertilization of F_1 individuals.

fascicle, fascicled. A close cluster or bundle of flowers, leaves stems, or roots.

fenestrate. Perforated with; windowlike holes or translucent areas.

ferruginous. Rust-colored.

fetid. Having a foul odor.

fibril. Submicroscopic threads or fibers that constitute the form in which cellulose occurs in the cell wall.

filament. The stalk of a stamen that supports the anther.

filiform. Threadlike, usually flexuous.

flabellate, flabelliform. Fan-shaped; broadly wedge-shaped.

flexuous. Coarsely sinuous or wavy; curved alternately in opposite directions.

floral bract. Structure just below the flower. In *Tillandsia*, they are often colorful.

floral envelope. The perianth, or calyx, and corolla.

floral tube. A tube or cup formed by the fused basal portions of the sepals, petals, and stamens that surrounds the ovary (fused or free from it); often in perigynous and epigynous flowers.

foliaceous. Leaflike; resembling a leaf in shape, texture, size, and/or color.

foliage. A collective term for the leaves of a plant.

foliar. Pertaining to leaves or leaflike parts.

founder principle. The principle that propagules that initiate a new and isolated population will contain less genetic diversity than the parent population from which they originated.

frass. The excrement of insects and larvae.

free. Separate, distinct; not united in any way to other organs or structures.

fulvous. Tawny; dull yellowish-brown.

furfuraceous, furfurate. Covered with bran-like scales; scurfy.

fuscous. Grayish-brown; dull brown.

fusiform. Spindle-shaped; thickest near the middle and tapering toward each end.

gamete. A mature, haploid reproductive cell or nucleus that unites with another gamete to form a zygote in sexual reproduction. *Syn*. **sex cells**.

gamo-. A prefix meaning "united."

GARDNER, CECELIA SUE (1939-). Taxonomist and botanist from Corpus Christi, Texas, who has specialized in the study of *Tillandsia*.

gas exchange. The inflow or outflow of oxygen and carbon dioxide during plant respiration.

generative nucleus. The nucleus in an immature pollen grain that divides mitotically to form two sperm nuclei.

genetic drift. In a small, isolated population, the chance elimination of a gene from the genotype due to the "fixing" of the opposite gene in that allele.

genotype. The genetic constitution of an individual, in contrast to its appearance or phenotype.

geotropism. A growth movement of a plant in response to the influence of gravity.

germ pore. Thin areas in the surface of a pollen grain.

GILMARTIN, AMY JEAN (1938-). Botanist, taxonomist, and professor at Washington State University who has done extensive work with the Bromeliaceae of Ecuador and bromeliad chromosome research.

glabrous. Without pubescence; not hairy. Smooth.

glaucous. Covered with a removable waxy coating that gives the surface a whitish or bluish cast.

GLAZIOU, AUGUSTE (1826-1906). French botanist, landscape architect, and horticultural adviser to Dom Pedro of Brazil, for whom he collected plants. From 1858-1871 he discovered more than 60 new bromeliad species.

globose, globular. Spherical or rounded.

glomerate. In dense or compact clusters or heads.

glucose. A common six-carbon sugar ($C_6H_{12}O_6$).

grex. A group of species or hybrids; applied collectively to the offspring of a given cross from one seed pod.

guard cells. Specialized crescent-shaped, chloroplast-containing cells that surround a stoma. Changes in the turgor of the guard cells result in the opening and closing of the stoma.

Guzmania. Epiphytic bromeliad genus that occurs in rainforests in parts of Central America and the Andean highlands of Colombia and Ecuador. Described by Ruiz and Pavon in 1802 for Guzman, an 18th century Spanish naturalist.

habit. General appearance or characteristic form of a plant.

habitat. Natural environment of an organism.

hair. *See* **trichome**.

haploid. Used in reference to either a nucleus, a cell, or an entire organism in which only one member of each set of homologous chromosomes is present. *Compare* **diploid**.

Hectia. Terrestrial bromeliad genus described by Klotsch in 1835 for Julius Hecht. Native to Texas, Mexico, and northern Central America.

herb. A nonwoody plant (annual, biennial, or perennial) whose aerial portion naturally dies to the ground at the end of the growing season.

heritability. Proportion of observed variability due to the cumulative effect of genes; the remainder being due to the environment.

hetero-. Prefix meaning "other" or "different."

heterogeneous. Not uniform. Differing.

heterotroph. An organism that cannot produce its own food and must consume other organisms or their organic products for food, e.g. animals, saprophytes, and parasites.

hoary. Covered with a white or grayish-white pubescence. *Syn*. **canescent**.

holotype. The single, type specimen, designated to carry the name of a new species.

homogeneous. Uniform, alike; having the same nature or consistency.

homologous, homology. Having the same phylogenetic or evolutionary origin, but not necessarily the same structure and/or function.

homologous chromosomes. Morphologically similar chromosomes that associate in pairs during prophase I of meiosis; each member of a pair is derived from a different parent.

homonym. In nomenclature, a name rejected because it duplicates a name previously and validly published for a group of the same taxonomic rank and based on a different type.

hyaline. Thin and translucent or transparent, like glass.

hyper-. Prefix meaning "above" or "over."

hypo-. Prefix meaning "under" or "less."

hypodermis. Layer or layers of cells beneath the epidermis, which are morphologically distinct from the underlying cortical or mesophyll cells.

hypogynous. Having a floral organization in which the flower parts are attached below the base of the ovary and free from it; flowers with this arrangement have a superior ovary. *Compare* **epigynous** and **perigynous**.

imbricate. With margins of structures overlapping like shingles on a roof.

included. Not protruding beyond the surrounding structure. *Compare* **exserted**.

incomplete dominance. Condition that results when two different alleles acting together produce a result intermediate between the effects of these same genes in the homozygous condition.

indeterminate. Growing from the apex, particularly the main axis. An inflorescence that continues to develop new flowers at the tip for an indefinite length of time.

indigen, indigenous. Native to a region or country; not introduced or imported. *Syn*. **native**. *Compare* **endemic**.

indument, indumentum. Any covering of a plant surface, especially pubescence.

inferior ovary. Ovary that is situated below the point of insertion of the other floral organs.

inflated. Puffed out, bladderlike.

inflorescence. Arrangement of flowers on a floral axis; a floral cluster.

internode. Portion of the stem between two successive nodes.

introgressive hybridization. Genetic modification of one species by another through the intermediacy of hybrids; introgression.

involute. Margins rolled or turned in over the upper or ventral surface.

ion. An atom or group of atoms with a negative or positive charge.

isolation. Separation of one population from another so that interbreeding is prevented; isolation may result from one or more factors such as geographical isolation or physiological isolation.

lanceolate. Lance-shaped; much longer than broad; widened above the base and tapering toward the apex.

lax. Loose, not densely packed. *Contrast* **congested**.

leaf axil. Upper angle between a leaf or leaf petiole and the stem to which it is attached.

lepidote. Covered with small, scurfy scales.

lignin. Complex organic substance derived from phenylopropane and distinct from carbohydrates. It is an important ingredient of the material surrounding the cellulose microfibrils in the secondary wall of many plant cells. It imparts rigidity and strength especially to woody tissues.

light reaction process. Reactions of photosynthesis in which light energy is required for activation.

ligulate. Strap or tongue-shaped.

ligule. Tiny, tonguelike flaps of tissue on the adaxial petal base in some species of bromeliad. Used to separate the genera of *Vriesea* and *Tillandsia*.

locule. A compartment, chamber, or cavity as in an anther, ovary, or fruit.

locus, *pl*. **loci**. Position of a gene on a chromosome.

lumen. Space within a plant cell bounded by the cell wall; the cavity left within a cell after the protoplast dies, as in trichomes.

LUTHER, HARRY (1952-). Taxonomist and director of the Bromeliad Identification Center at the Marie Selby Identification Center in Sarasota, Florida.

MARNIER-LAPOSTOLLE, JULIEN (1889-1976). Horticulturist, collector, creator of Les Cedres, the famous botanical garden in southern France; contains the world's largest private collection of bromeliads.

McWILLIAMS, ED (1941-). Horticultural professor, botanist, and taxonomist at Texas A & M University who has worked with bromeliads.

meristem. The areas that produce new leaf, root, or floral cells, i.e. tissue in which mitotic cell divisions occur.

mesic. Moist, especially in reference to habitats or environments. *Compare* **xeric**.

mesophyll. Photosynthetic tissue between the upper and lower epidermis of a leaf.

mitochondrion, *pl*. **mitochondria**. A double-membrane-bound, cytoplasmic organelle of eukaryotic (organisms that have true nuclei, other membrane-bound organelles, chromosomes, etc.) plants, in which cellular respiration occurs.

monocarpic. A perennial or annual blooming and fruiting only once and then dying.

monocotyledon. A member of the Monocotyledoneae, one of the two taxa of angiosperms, characterized by the following: one seed leaf, flower parts usually in threes (or multiples thereof), leaves often with parallel venation, root systems arising adventitiously.

monoecious. A plant with separate male and female reproductive structure occurring on the same plant. In angiosperms, having unisexual flowers of different sexes on the same plant.

monograph. A comprehensive treatise on a single taxonomic unit, such as a family or genus.

MORREN, CHARLES JACQUES (1807-1858). Founder and editor of *La Belgique Horticole* and authority on bromeliads. At the time of his death was the leading champion of Bromeliaceae. Director the Botanic Garden at Liege and a well-known color illustrator.

mucronate. Terminated abruptly by a distinct, short, sharp point.

myrmecophily. Pollination by ants.

myrmecophyte. Plant adapted to live in symbiosis with ants and/or ant colonies.

natural selection. Process of evolution that involves the interaction between an organism and its environment resulting in a differential rate of survival and reproduction of different genotypes in a population. The individuals best adapted to their environment are favored (selected) over the less adapted.

nerved. An easily visible vein in a leaf or other structure.

***Neoregelia*.** Bromeliad genus mostly native to Brazil. Species are usually characterized by flat rosettes that develop colorful centers when in bloom. Described by Lyman B. Smith in 1955 for the Russian Botonist Edward von Regel.

neutrality. Physical characteristics that are neither adaptive nor nonadaptive.

***Nidularium*.** Bromeliad genus confined to eastern Brazil. Named by Lemaire in 1836 from the Latin word for "nest," a reference to the flowers that emerge from among the center leaves.

node. A point along a stem where one or more leaves or bracts are attached.

nucleus. A large organelle containing most of the genetic material in a cell.

obicular. More or less round or circular in outline or shape.

offset. A short lateral shoot that develops from the main stem of a plant, providing a means of asexual reproduction. A pup or offshoot.

oligotrope. An organism adapted to grow in nutrient-deficient habitats.

organelle. A subcellular body located in the cytoplasm, e.g. mitochondria, chloroplast, etc.

orthotropic. Growing vertically (toward light).

osmosis. Diffusion of a solvent (usually water) through a differentially permeable membrane, from the side containing the higher concentration (of water) to the side containing the lower concentration. The migration continues until water concentration is equal on both sides.

osmotic potential (pi, π). That component of water potential (psi, ψ) that is due to dissolved substances.

ovary. The hollow basal part of the pistil that contains the ovules.

ovate. Egg-shaped in outline, the broader end is downward.

ovoid. A solid object that is oval or egg-shaped in outline and the broader end is downward.

ovule. Structure in which the female halophase will develop—should the resulting egg in the embryo sac be fertilized, the ovule then becomes a seed.

PADILLA, VICTORIA. Author of several books, collector, lecturer. Founding member of the Bromeliad Society, Inc. Editor of the Journal of the Bromeliad Society magazine and driving force of the Society from 1960 to 1980.

parenchyma. The most common type of plant cell, it is usualy thin-walled and concerned with one or more physiological activity, e.g. photosynthesis, movement of water and food, storage, etc.

pectic compounds (pectin). Certain chemical substances, especially abundant in the middle lamella, that bind adjacent cells together.

pectin. *See* **pectic compounds**. Pectin is a pectic compound; the basis of fruit jellies.

peltate. Umbrella or shield-shaped, with a stalk attached to the lower surface near the center of the structure, as in most bromeliad trichomes.

perforation plate. The walls at each end of a vessel element that are perforated by one or more openings.

perianth. Collective term for the floral envelope as a whole, consisting of the calyx (sepals) and corolla (petals).

perigynous. Flower whose ovary is enclosed by but not fused to a floral tube. *Compare* **epigynous** and **hypogynous.**

petal claw. The long, narrow base of the petals or sepals in some flowers.

petal scales (flaps). Tiny flaps of tissue present on the inner surface of each petal in some bromeliad species including all vrieseas.

pH value. Chemical symbol denoting the relative hydrogen ion concentration in a solution or soil, i.e. a measure of the acidity or akalinity. A pH value of 7.0 is neutral, less than 7.0 acidic, over 7.0 alkaline.

phalaenophily. Pollination by moths.

phanerogam. Plant that reproduces by seed. *Compare* **cryptogam.**

phenetic. Classification system based on current, total simularity, using as many separate characteristics as possible, each of which is given equal weight; e.g. morphological, physiological, etc. Does not reflect phylogeny because the evolutionary history is unknown.

phenology. Relationship between an organism's activities and the seasonal calendar.

phenotype. External physical appearance of an organism resulting from interaction between the genetic constitution (genotype) of the individual and the environment.

phloem. Food-transporting tissue consisting of several parenchyma cell types including sieve tube elements and companion cells. Also fibers and sclereids.

photobleach. Overexposure of leaf cells to light causing damage to chlorophyll. Leaves generally become more yellow and less green.

photosynthesis. Process whereby light is converted into chemical energy in the presence of chlorophyll. It involves the production of a carbohydrate from carbon dioxide and water with the release of oxygen.

photosynthetically active radiation (PAR). The 400-700 nanometer (10-9 m) wavelength portion of the sun's radiant energy used by green plants to make food.

phyllotaxis. Arrangement of leaves or bracts on a stem.

phylogenetic, phylogeny. Pertaining to the evolutionary history and relationships among a group of organisms.

phytophagus. Herbivorous.

pinnate. Formed like a feather, i.e. having the leaflets of a compound leaf placed on either side of the rachis.

pistil. Female apparatus of most flowers, consisting of a stigma, style, and ovary.

pit. Thin spot in the wall of a plant cell. Simple pits have parallel sides; bordered pits have overhanging sides.

plasma membrane. Differentially permeable membrane bounding the protoplast of a cell.

plasmolysis. Separation of the cytoplasm from the cell wall caused by the removal of water from its protoplast by osmosis.

pleitropy. Capacity of a gene to affect a number of different phenotypic characters.

plicate. Folded like a fan, with a number of longitudinal parallel ridges.

plumose. Covered with fine hairs, featherlike pubescence.

pollen. Spores or grains borne by the anther containing the male element.

pollen tube. An outgrowth from a germinating pollen grain that conveys a sperm down the style to an ovule.

polymer. Large molecule such as cellulose or a protein composed of linked smaller subunits that are chemically identical.

polyploid. A cell, tissue, or organism whose cell nuclei contain more than two complete sets of chromosomes.

polystichous. Stems bearing leaves or bracts attached in a circular or spiral arrangement; having many vertical ranks.

population. Any group of organisms of the same species living in a particular area at the same time.

primary bract. Structure at the base of each bloom spike; present only in a compound inflorescence, often leaflike but may be colored.

proboscis. Extendable tubular structure used for sucking by some insects and mollusks.

proliferation. Producing offsets; growing by multiple divisions.

propagule. Any structure, sexual or vegetative, that becomes separated from the parent plant and then serves as a continuing means of propagation.

protein. Complex organic compound consisting of amino acids and connected by peptide bonds.

protoplasm. Inclusive term for all living material of a cell.

protoplast. A living cell minus its cell wall.

pup. *See* **offset**.

putative. Hypothesized.

Puya. Bromeliad genus from the high Andes. Considered the oldest in the family. Described by Molina in 1782, *puya* is an Indian word for "point."

rachis. Major axis of a compound leaf or inflorescence.

rainforest. Special, closed plant community dominated by trees that form a two or more layered, dense canopy in which lianas and epiphytes are most conspicuous, and a lower sparse assemblage of smaller trees, shrubs, herbs, and ferns.

READ, ROBERT (1931-). A curator with the National Herbarium at the Smithsonian Institute in Washington, D.C., who has worked extensively with the taxonomy of Bromeliaceae.

receptacle. The portion of the axis of a flower stalk on which a flower is borne.

recurved. Bent or curved downward or backward.

reflexed. Abruptly recurved downward or backward.

refugia. Isolated ecosystems.

respiration. Intracellular series of enzyme-mediated chemical reactions in which food is oxidized and energy is released. The complete breakdown of the organic compounds also produces carbon dioxide and water.

RICHTER, WALTER EMIL (1904-). Eminent German bromeliad grower, author, and hybridist. Since 1930, he has produced outstanding bromeliad crosses.

ring cells. Circular arrangement of cells located between the four central cells and the wing of a tillandsioid trichome.

rosette. *adj.* **rosulate**. Crowded, circular cluster of leaves or other organs; here used in reference to a growth habit in which leaves radiate from a crown.

saphrophyte. Plant that obtains its food from decaying organic matter.

saxicolous. Habit of growing on rocks.

scape. Stem of the inflorescence usually extending beyond the leaves.

scape bract. Structures borne on the scape. They may be leaflike or colored.

sclerified. Having thick, lignified cell walls.

secund. Arrangement of leaves or floral bracts growing on one side of an inflorescence.

sepal. One of the outermost, sterile appendages of a flower that normally encloses the other floral parts in the bud; one of the separate parts of the calyx.

septicidal. Dehiscence along or in partitions, not directly into the locule.

sheath. More or less leaflike structure surrounding an organ or plant part.

shield. Plate of dead cells attached to the top of the stalk of a peltate trichome.

shoot. Collective term that includes all aerial portions of a vascular plant. Synonym for an individual plant.

sieve plate. Perforated areas at each end of sieve tube element. Individual food-conducting cell making up the sieve tubes of phloem tissue.

simple. Of one piece; not compound; a single unbranched inflorescence; an inflorescence with a single flower spike.

sink. A part of the plant where more food is required than that region can produce.

solute. The dissolved substance—gas, solid, or liquid—in a solution.

solvent. Medium in which a substance is dissolved.

source. That part of a plant that exports food to regions where it is required (sinks).

speciation. The evolutionary process by which new species are formed.

spectrum. Range of visible colored bands diffracted and arranged in order of their respective wavelengths from longest (red) to shortest (violet) as a result of white light being passed through a prism or other defracting medium. Can also include invisible components at both ends of the spectrum, i.e. infrared and ultraviolet.

sphagnum. A type of soft moss in the genus *Sphagnum*, that is used for decorative purposes in the floral industry. Specifically, often used for cosmetic purpose in the mounting of tillandsias.

spike. Unbranched, indeterminate, elongated inflorescence that bears sessile flowers.

spreading. Diverging outward, almost horizontal.

stalk cells. Cells of the trichome directly beneath the shield that penetrate the epidermis. The handle portion of the peltate scale (trichome).

stamen. Male appendage of the flower consisting of the filament and anther.

stolon. Prostrate stem that forms a new plant at its terminus.

stoma, *pl.* **stomata**. Minute pore in the epidermis of a leaf, stem, or reproductive structure that functions in gaseous exchange between a plant and its external environment. Stoma are bordered by guard cells that function in the regulation of the size of the opening.

stramineous. Straw-colored.

striate. Having fine, longitudinal lines, grooves, streaks, or channels arranged more or less parallel.

strict. Stiff, rigid, erect.

stroma. Liquid matrix within a chloroplast where many of the enzymes involved in photosynthesis are located.

style. That portion of the stigma located between the stigma and the ovary.

sub-. Prefix meaning "under," "below," or "almost."

subgenus. One of the divisions into which large genera are sometimes taxonomically divided.

substrate. The surface or medium on or in which an organism is living and from which it gets its nourishment.

subtend. Occurring immediately below or near, as in sepals subtending petals.

subulate. Awl-shaped, tapering from base to apex.

succulent. 1. Plant that accumulates water in fleshy, water-storing stems, leaves, or roots. 2. Juicy, fleshy, in reference to texture or appearance.

sucrose. Common table sugar; a disaccharide (glucose plus fructose) found in many plants. It is the primary form in which sugar, produced as a result of photosynthesis, is translocated in a plant.

symbiosis. An association between two different kinds of organisms that benefits both parties.

sympatric. 1. Originating in or occupying the same geographical region. 2. Species or populations that occur close enough together to be within the range of mutual pollinating vectors.

synonym. Scientific name rejected in favor of an accepted correct name for a species of plant.

synonymy. Complete listing of the scientific names applied to a particular taxon. Usually appears with other data, e.g. date of publication, author of publication, etc.

syntax. The arrangement of words (in their proper forms) by which their connection and relation in a sentence are shown.

systematics. Botanical taxonomy; the study of the relationships among plants.

taxon, *pl.* **taxa**. Any taxonomic unit into which living organisms are classified, e.g. family, genus, species.

taxonomy. Science of the classification of organisms.

terrestrial. Growing in the ground and supported by soil as opposed to growing in trees or water.

thigmotropic. Growth or movement in response to a mechanical stimulus, e.g. the leaves of a Venus Flytrap moving in response to touch.

thylakoid. Flat, sacklike membranous subunit of the chloroplast. Stacks of thylakoids form the grana.

tomentum. Covering of woolly, densely matted hairs.

tonoplast. A one-layered, cytoplasmic membrane bounding the vacuole of a cell.

tracheids. Elongated, thick-walled, nonliving conducting and supporting cell found in the xylem.

transpiration. Evaporation of moisture through the surface of a plant.

trichome (hair, scale). Epidermal structure.

turgor pressure (**P**). Pressure exerted against a cell wall as a result of movement of water into the cell.

type (specimen). *See* **holotype**.

UTLEY, JOHN & KATHLEEN (1944-). Botanists, professors, taxonomists at the University of New Orleans who have specialized in the study of *Vriesea* and *Hectia*.

vacuole. Membrane-bound, water-filled vessicle within the cytoplasm of plant cells; they are filled with a watery liquid, the cell sap, rather than with protoplasm. Vacuoles function in maintenance of water balance within the cell, accumulation of various ions and molecules, storage of pigments, etc.

vascular. Refers to any plant tissue or area that consists of or gives rise to conducting tissue, e.g. xylem, phloem, or vascular cambium.

vascular bundle. Strand of vascular tissue composed of xylem phloem.

ventral. Pertaining to the surface nearest the axis; the upper surface of a leaf; the inner surface of an organ. *Syn.* **adaxial**. *Ant.* **dorsal**.

vesicle. Small sac or cavity; a spherical body.

vessel element. Advanced type of vascular xylem tissue.

viviparous. Germinating while still attached to the parent plant.

water potential (psi, ψ). A measure of the free energy status of water that determines the direction of water movement in the plant body from areas with higher to those with lower water potentials.

WEBER, WILHELM (1916-1986). Bromeliad enthusiast from the German Democratic Republic who developed his keen taxonomic interest during a career as a perfume specialist. He described many new bromeliad species over the past 10 years—an incredible feat for one not able to travel or explore.

whorl. Three or more leaves or flowers at one node; a circle of floral organs, e.g. stamens, sepals, petals, etc.

wing. Outermost portion of a tillandsioid trichome shield, peripheral to the ring cells.

xeric. Pertaining to dry places; adapted to arid conditons.

xerophyte. Plant adapted to withstand drought.

xylem. Principle water- and mineral-conducting tissue in vascular plants; a complex tissue composed of nonliving, lignified tracheids, vessels, and fibers, and their associated parenchyma cells.

Index

For major topics, the most important page numbers are starred (*).

Illustrations (photographs or figures) in the text are indicated by page numbers in italics.

Names of persons discussed in the text or cited in references to the literature are indexed. Plant species are indexed wherever they appear in the text. Terms defined in the text are indexed, but the Index does not cover the Glossary.

Index compiled by Barbara Wurf